Plant Taxonomy and Biosystematics

Second edition

Clive A. Stace

Professor of Plant Taxonomy, University of Leicester

CAMBRIDGE
UNIVERSITY PRESS

PUBLISHED BY THE PRESS SYNDICATE OF THE UNIVERSITY OF CAMBRIDGE
The Pitt Building, Trumpington Street, Cambridge, United Kingdom

CAMBRIDGE UNIVERSITY PRESS
The Edinburgh Building, Cambridge CB2 2RU, UK   http://www.cup.cam.ac.uk
40 West 20th Street, New York, NY 10011–4211, USA   http://www.cup.org
10 Stamford Road, Oakleigh, Melbourne 3166, Australia
Ruiz de Alarcón 13, 28014 Madrid, Spain

First edition published by Edward Arnold 1980
Second edition first published by Edward Arnold 1989 and first published by
Cambridge University Press 1991
Reprinted 1993, 1994, 1996, 2000

Printed in the United Kingdom at the University Press, Cambridge

British Library Cataloguing in Publication data
Stace, Clive A. (Clive Anthony), *1938–*
    Plant taxonomy and biosystematics.——2nd ed.
    1. Plants.   Taxonomy
    I. Title
    581'.012

Library of Congress Cataloguing in Publication data available

ISBN 0 521 42785 1 paperback

Contents

Preface to first edition

The aim of this book is to explain in general terms the nature of plant taxonomy, using, where appropriate, particular examples but without presupposing any knowledge of systematics. It is designed to present taxonomy as a contemporary science by describing the current aspirations of taxonomists and the principles and methods which underlie them.

The book is primarily intended to offer a text for university undergraduates and others in tertiary educational establishments, but it is hoped that serious amateurs, teachers and research workers who specialize in other fields but utilize taxonomic information will find it useful. The majority of other books on plant taxonomy are either less concise or more specialized; many are now considerably out of date and many devote a good number of pages to a systematic treatment of plant families. Two important general works should be mentioned: *Principles of Angiosperm Taxonomy* (Davis and Heywood, 1963) and *Vascular Plant Systematics* (Radford *et al.*, 1974). Both contain much valuable information, the first stressing general principles and the second acting as a source-book for taxonomic data and methodology. Both of these, however, are far larger and contain much more detailed texts than the present one, and they serve different purposes.

Despite its title, the subject matter of this book is taxonomy. Topics discussed in relation to an understanding of modern plant taxonomy, such as historical taxonomy, cytogenetics, structural botany, biosystematics and evolutionary theory, are dealt with not for their own sakes but in an attempt to clarify the all-embracing nature of taxonomy. In particular, biosystematic data have been woven into the text in what is regarded as their rightful places, not segregated into separate chapters or sections; the title of the book is intended to underline this feature. Such an arrangement inevitably excludes many aspects of plant science (such as incompatibility mechanisms, population dynamics, micro-evolutionary processes, physiological ecology) which are often covered by the term biosystematics, but which belong to the fields of ecology, physiology, genetics, population biology or phylogenetic investigations rather than to taxonomy; their omission has been deliberate.

Naturally, much of what is discussed in this book is equally true of the taxonomy of animals or microorganisms; but much is not, because these organisms do show fundamental differences from plants (as well as fundamental similarities to them). For that reason this book is largely limited to a consideration of green plants; fungi (and lichens) are not included, as there now seem to be good biochemical and structural reasons to believe that they should be classified separately. Within this ambit examples have been drawn from all groups, from algae to angiosperms, but the latter are mentioned most frequently because they are the most numerous, familiar and well investigated. In order to place this taxonomic delimitation into perspective, and to supply a readily available list of the major taxonomic groups (both scientific and vernacular) mentioned in the text, an outline classification of plants is provided in the Appendix. I cannot defend every aspect of this classification against contrasting features in other classifications, but the system adopted appears reasonably in accordance with most modern views and it is adhered to throughout this book for the sake of conformity and clarity.

One other feature of the classification adopted must be mentioned. The usual termination for plant families, as laid down in the *International Code of Botanical Nomenclature*, is *-aceae*, but in the angiosperms eight exceptions are made for family names of very long standing and great familiarity; in these cases two alternative, equally valid names are permitted. In this book the names ending in *-aceae* are used exclusively, a choice which is becoming increasingly popular throughout the world. The names used here, with their older equivalents, are: Apiaceae (Umbelliferae), Arecaceae (Palmae), Asteraceae (Compositae), Brassicaceae (Cruciferae), Clusiaceae (Guttiferae), Fabaceae (Leguminosae), Lamiaceae (Labiatae) and Poaceae (Gramineae).

The plan for this book has gradually evolved over the past fifteen years from undergraduate courses given wholly or jointly by me in the Universities of Manchester and Leicester, and it owes a great deal to many biologists there and elsewhere who have assisted me by tuition, argument, criticism, discussion and advice on a multitude of topics. Nevertheless, the views and opinions expressed here are, unless otherwise attributed, my own, and for these and for any errors of fact I accept full responsibility.

Leicester   C. A. Stace
1979

Preface to second edition

Since the first edition in 1980, two major developments have had a noticeable impact on plant taxonomy: firstly, the general availability of molecular biological techniques; and secondly, the application of the principles of cladistics. In this edition I have therefore paid particular attention to these topics, which have necessitated reorganizing parts of Chapters 2, 4, 5, 7 and 8 to accommodate the extra material. I have also tried to update all the other chapters and the References. Today taxonomy is more exciting and challenging than it ever has been, and I hope that I have succeeded in conveying this ambience to the reader.

At the time of writing, the future role of cladistics in plant taxonomy is uncertain. Already the proportion of cladists that adheres to all aspects of the original thesis has diminished, as more and more of the basic tenets propounded by Hennig are compromised in the light of experience. My own views are both positive and negative. I am strongly *for* a close examination of the principles and methods of cladistics in order to find how the subject can most profitably be absorbed into taxonomic practice (with the expectation that it has much to offer), but strongly *against* acceptance of the frequently stated dogmatic views that it is always the best or is the only worthwhile line of phylogenetic investigation, or that phylogenetic classifications are *necessarily* better (more natural or predictive) than phenetic ones.

Many aspects of the classification outlined in the Appendix could be improved upon, but classification of the upper levels of the plant kingdom is today as fluid as ever, and I have resisted the temptation for change in the interests of continuity and clarity.

During the preparation of this Second Edition I have had the benefit of discussion with numerous colleagues. Parts of Chapters 2 and 7 have been kindly commented upon by Dr R. J. Gornall, and of Chapter 4 by Dr J. Draper. Professor A. J. Willis (Sheffield) read the manuscripts of both editions and made numerous helpful comments and criticisms. I am very grateful for all their help, and also to those who wrote to me concerning the first edition with views, queries and criticisms.

C. A. Stace
Leicester
1988

Section 1
The Basis of Plant Taxonomy

Introduction

Taxonomy can lay claims to being the oldest, the most basic and the most all-embracing of the biological sciences; it is certainly one of the most controversial, misunderstood and maligned. These properties are all closely related to the nature of taxonomy itself. As one of its purposes is to provide a service to non-taxonomists, its principles and practices come under scrutiny by non-specialists more often than is the case with most sciences, and the lack of understanding resulting from such usage is the cause of much of the mistrust and criticism often directed at it.

But this situation is partly the fault of taxonomists themselves, for they have not yet fully succeeded in formulating common goals or in educating other biologists in taxonomic principles, despite the many volumes which have been written on the subject. In particular, they have not conveyed to a wide enough audience the concept of their science as one which does not have an obvious, finite, single aim or purpose, and as one which has no data of its own but uses those from every other field of biological investigation. Since the use made of these data must necessarily vary according to the precise aims and expertise of the investigator, there is a considerable element of subjectivity in taxonomic methodology, and it is difficult to see how it can be avoided. Attempts made in recent years to eliminate this subjectivity, mainly by the use of taxometrics and cladistics, have still had little impact on actual classifications, although they have been valuable in teaching taxonomists to analyse more carefully the nature of the data which they use and the way in which they use them.

The prominence of taxonomy in biology has changed much over the past century, with changing technology, tastes and requirements. From a pre-eminent position last century, taxonomy declined in popularity as new fields of investigation, such as genetics, electron microscopy and molecular biology, emerged. These new subjects appeared to many biologists more unifying, consequential and exciting than taxonomy, and they occupied the minds of a high proportion of biologists and their students.

Nevertheless, there comes a point in all fields of biological research when the initial emphasis on the universal application of the principles being uncovered changes to an emphasis on the interspecific differences, such as we are at present witnessing in many different spheres of biochemistry. The study of such differences can only be properly documented, and their significance fully grasped, by the employment of taxonomic expertise and, later, by their expression as systematic information within taxonomic classifications. This enigmatic aspect of taxonomy is one of its fundamental properties; it is at the same time the most basic and the ultimate (the most derived or synthetic) of all the biological sciences. It is basic because no start on understanding the wealth of variation can be made until some sort of classification is adopted, and ultimate because taxonomy is not complete until the data from all other fields of investigation have been incorporated. This dual property is not, of course, manifested in two separate stages, but by a continuous process of anabolism and catabolism as more and more data are utilized.

At present it seems reasonable to claim that the decline of taxonomy, evident for much of this century despite the periodic injection of new ideas and tools, has been slowed or even halted, and there is an increasing acceptance of the fundamental importance of taxonomy and the need for a firmly based, comprehensible methodology associated with it. It is vital that taxonomy is taught as the broad, challenging and exciting subject that taxonomists know it to be, and that this tuition is carried out at the undergraduate level in its proper place alongside the other biological disciplines.

Studied and taught in this context, taxonomy will be seen as a subject with broad horizons and far-reaching consequences, and taxonomists as scientists with a key role to play in biological research, conservation and food production. It is to be hoped that the notion that taxonomy is very largely subjective and intuitive, an art as much as a science, will become generally abandoned without losing sight of the fact that an element of subjectivity is inevitable and a degree of intuition desirable.

The idea that the study of taxonomy must necessarily involve the amassing of an encyclopaedic knowledge of plant diversity has long deterred would-be taxonomists. Indeed, at one time university courses (and textbooks) in taxonomy did pay a great deal of attention to this aspect. Of course, the ability to retain a lot of information of this sort is a great advantage, just as (for example) a biochemist would be aided if he could remember the complete chart of metabolic pathways, but the concept of the subject being mainly concerned with attaining such a knowledge is totally untenable. Some taxonomists, particularly those working in herbaria and providing an identification service, do need such specialized expertise. However, in teaching taxonomy to undergraduates, most of whom will not become taxonomists, or to experts in other fields, it is important to illustrate the aims, the principles and the methods rather than this or that flora, and to explain how to understand and utilize taxonomic literature and the information which it contains rather than how to memorize it.

1
The scope of taxonomy

Taxonomy may be defined as the study and description of the variation of organisms, the investigation of the causes and consequences of this variation, and the manipulation of the data obtained to produce a system of classification. Such a definition is wider than that sometimes given, and has intentionally been drawn up to coincide with the meaning of the term *systematics*. In fact the two terms are nowadays commonly used synonymously, and are so-treated in this book. It should, however, be realized that some authors prefer to differentiate between them, in which case systematics has more or less the broad definition given above, and taxonomy is restricted to the study of classification.

Classification (as a process) is the production of a logical system of categories, each containing any number of organisms, which allows easier reference to its components (kinds of organisms). Classification (as an object) is that system itself, of which there are many sorts.

Identification or *determination* is the naming of an organism by reference to an already existent classification. The term classification is often wrongly and misleadingly used in this sense; this is to be discouraged, for classification must necessarily precede determination.

The study of the system and methods of naming organisms, and the construction, interpretation and application of the regulations governing this system, are covered by the term *nomenclature*.

A *taxon* (pl. taxa) is any taxonomic grouping, such as a phylum, a family or a species. It is a useful general term, and can be used to indicate the rank of a group as well as the organisms contained within that group.

Taxa are delimited by various means, but the fact that they can be delimited at all (at any level) illustrates an essential feature of most taxonomic variation—that it is *discontinuous*, and that taxa are circumscribed by their discontinuities from related taxa.

A *description* of a taxon is a statement of its characteristics, which thus constitute the definition of that taxon. Characters contributing to a taxonomic description are known as *taxonomic* or *systematic characters*. A *diagnosis* is a shortened description covering only those characters (*diagnostic characters*) which are necessary to distinguish a taxon from other related taxa.

A *flora* (lower case initial letter) is the plant life of any given area. A *Flora* (upper case initial letter) is a book or other work describing the flora of a given area, and usually providing a means of determining the taxa contained in it. *Floristics* is the study of floras, including the preparation of Floras.

Attempts are often made to differentiate between different facets or lines of approach to taxonomy; in particular, a distinction might be made between *experimental taxonomy* or *biosystematics* and *orthodox* or *classical taxonomy*. Experimental taxonomy does not simply imply the use of experimental procedures, but is the taxonomic study of organisms from the standpoint of populations rather than individuals, and of the evolutionary processes which occur within populations. Hence the term biosystematics is preferable. It is, inevitably, largely concerned with genetical, cytological and ecological aspects of taxonomy and must involve studies in the field and experimental garden, whereas orthodox taxonomy more often relies on morphological and anatomical data and can be carried out to a large degree in the herbarium and laboratory. Biosystematics may therefore be considered as the taxonomic application of the discipline known as *genecology*—the study of the genotypic and phenotypic variation of species in relation to the environments in which they occur. It is unfortunate that the term biosystematics has been widened by some taxonomists to cover virtually any taxonomic activity not pursued in the herbarium or almost any newly acquired technique. It should be emphasized that it is not the nature of the data used, be they morphological, cytological or chemical, but the use to which they are put which differentiates between classical taxonomy and biosystematics. However, it must be equally stressed that these two fields are not separate and opposing, but rather are closely interacting, complementary approaches to taxonomy, without either of which taxonomy is incomplete.

Other distinctive taxonomic activities which have been considered worthy of a name are narrower and more easily defined. Hence chemical taxonomy is the study and use of chemical characters in taxonomy; systematic anatomy is the study of anatomy with a view to extracting taxonomic data; and numerical taxonomy is the treatment of taxonomic data by numerical, normally computerized, methods; and so on.

The need for classification

The need for some system of classification is absolute, for it is only by first naming organisms and then grouping them in recognizable categories that one can begin to sort out and understand the vast array which exists. This requirement is not confined to taxonomists, or even to biologists, for living organisms are part of the every day life of all humans. Thus it is not surprising that classification is a process which mankind naturally and instinctively carries out, and which has been carried out from the very beginning, for the accurate recognition (identification) of food, predators, mates, fuel, building materials, etc. is essential for his survival.

Consideration of the numbers of kinds of plants known shows that the problem of classifying them all is enormous. Almost 300 000 species of green plants are currently recognized, plus over 100 000 fungi and a few thousand bacteria and other microscopic organisms which some biologists would classify as plants (Table 1.1).

Table 1.1 Estimated numbers of species of organism in the world. Data compiled from various sources, especially Prance.[329]

Seed plants	240 000
Pteridophytes	12 000
Bryophytes	23 000
Algae (eucaryotic)	17 000
Fungi	120 000
Lichens	16 500
Blue-green algae	500
Bacteria	3 000
Protozoa	30 000
Non-chordate animals	1 000 000
Chordate animals	50 000
	1 512 000

In 1938 Turrill[442] estimated that around 2 000 new species of flowering plants were described annually; today the number is approximately the same (11 538 new species were described between 1981 and 1985; figure supplied by Royal Botanic Gardens, Kew). It is clearly quite impossible for any one botanist to know more than a tiny fraction of the total number, but if they are grouped into larger units one can ascertain that an unknown plant belongs to, say, the Chlorophyta, or to the Rosaceae, both of which are taxa containing many species of which some are likely to be familiar.

In practice, the grouping of species into one of several larger units is not sufficiently precise; there are, for instance, over 5 000 species of Chlorophyta and over 2 000 species of Rosaceae. It is thus even more informative to learn that the unknown plant belongs to, say, the Volvocales, a subgroup of the Chlorophyta, or to the Potentilleae, a subgroup of the Rosaceae. For this reason the great majority of systems of classification combine species into an ascending series of successively larger and wider categories, ultimately arriving at a single all-embracing group covering all plants. Such systems are termed *hierarchical classifications* or *hierarchies*. These may be illustrated in two different ways, a plan-view or an elevation-view so to speak, both of which show how one or more taxa at each level are combined into a single taxon at the next higher level, this higher taxon being defined by the sum of the characters of all its subordinate taxa. The plan or bird's-eye view is often called the box-in-box presentation (Fig. 1.1B), while the elevation or side view is known as a *dendrogram* (Fig. 1.1A).

The hierarchy

Theoretically there is no limit to the number of levels contained in a hierarchy. In Fig. 1.1 five levels or *ranks* (species, genus, family, division or phylum, kingdom) are represented, but many more are commonly employed. *The International Code of Botanical Nomenclature*, the internationally agreed rule-book of green plant (and fungal) nomenclature, recognizes twelve main ranks in the hierarchy (Kingdom, Division, Class, Order, Family, Tribe, Genus, Section, Series, Species, Variety, Form, see Table 1.2) and this number can be easily doubled by designating subcategories below each rank (subdivision, subgenus, subspecies, etc.). Exceptionally, supercategories may

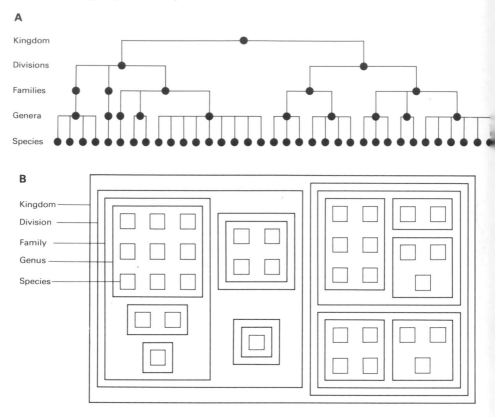

Fig. 1.1 Elevation (dendrogram, **A**) and plan (box-in-box, **B**) views of the same hierarchy. Note that the number of taxa in each successively higher taxon ranges from one to nine in this example. In **A** each taxon is represented by a spot, in **B** by a rectangle.

be designated above some of the ranks (e.g. superorder), and the *Code* also sanctions the interpositioning of additional ranks should they be thought necessary. In zoological and bacteriological nomenclature the hierarchy is substantially the same as in botanical nomenclature, except that in zoology the division is known as the phylum, a term which is also gaining favour with many botanists.

The names of these ranks should be used only in their strict senses. Hence we should not talk of divisions, subdivisions, classes, sections, series, varieties or forms in their loose, everyday contexts. Instead, taxonomically 'neutral' words such as group, taxon or variant should be employed where a particular rank is not implied. Thus it follows that the ranks must be used only in the precise sequence laid down by the *Code*, although, as indicated in Table 1.2, certain ranks may be omitted altogether if they are not needed. If these precautions are not taken, then the precision of the hierarchy and the enormous advantages that such an internationally accepted list of ranks bestows will be lost.

As an aid to indicating the ranks to which a taxon belongs, the names of taxa at ranks between division and subtribe are generally provided with

Table 1.2 The hierarchy of taxonomic ranks. The principal ranks are shown in capitals; the others may be omitted if not required. The example shows the classification of a dog-rose known as *Rosa canina* var. *lutetiana* f. *lasiostylis*.

Rank	Ending	Example
KINGDOM		Plantae
Subkingdom	-bionta	Embryobionta
DIVISION (Phylum)	-phyta	Tracheophyta
Subdivision	-phytina	Spermatophytina
CLASS	-opsida (-phyceae in algae)	Angiospermopsida
Subclass	-idae (-phycidae in algae)	Dicotyledonidae
(Superorder)		(Rosanae)
ORDER	-ales	Rosales
Suborder	-ineae	Rosineae
FAMILY	-aceae	Rosaceae
Subfamily	-oideae	Rosoideae
Tribe	-eae	Roseae
Subtribe	-inae	
GENUS		*Rosa*
Subgenus		*Rosa*
Section		*Caninae*
Subsection		*Caninae*
Series		
Subseries		
SPECIES		*canina*
Subspecies (subsp., or ssp.)		
Variety (var.)		*lutetiana*
Subvariety (subvar.)		
Form (f.)		*lasiostylis*
Subform (subf.)		

characteristic endings, as shown in Table 1.2. These endings are not the same in all groups of organisms, but within green plants the only important exceptions are found in the algae. Greater differences exist in the nomenclature of fungi and especially of animals, whereas in bacteriology the endings are based on those of plants. Rigid adherence to these recommendations can only serve as an aid to the acceptance and understanding of taxonomic principles, and is strongly advocated here.

The names of genera are nouns in the singular number, whereas the names of taxa at lower ranks may be nouns or adjectives in the singular or plural, and none is provided with distinctive endings denoting its precise rank. The rank should, therefore, be stated whenever it is not clear from the context. Further details concerning the taxonomic hierarchy are discussed in Chapters 8 and 9.

The concept of predictivity

The existence of such a well-defined and universally recognized hierarchical system is impressive, and at first might suggest that it must naturally have led to a high degree of unanimity concerning an ideal system of classification of plants. Perusal of a range of books, reviews and research articles shows, however, that this is far from the case. Even if outdated schemes of classification are set aside, there still exist many different systems, some differing only in detail but others in their broadest outlines, and each of these schemes has been adopted by a smaller or greater number of botanists. It is patently not the case that one of these schemes is correct, and the others wrong. Each, to a lesser or greater extent, has good and bad points. But it has not been found possible to incorporate all the good points of the various schemes into a single system of classification which also eliminates all the bad points. There remains an element of personal choice in the classification which is adopted, and the question still remains: how does one decide which is the best system to adopt? The plain answer to this is that it depends upon what is meant by the *best* system.

One could, for example, classify plants like inanimate objects—by size, shape, colour, texture, etc.—or even alphabetically if they had a name (as in the herbals of Fuchs and Turner, see p. 18), or in numerical sequence if they had a number. Systems of this sort are ideal if they are intended simply as a pigeon-holing exercise, whereby the classification aims only to enable users to find a plant in, say, a book, garden or herbarium as quickly as possible. Such systems, as well as the groups which they delimit, are termed *artificial*, and many of the very early schemes of classification were of this type.

The best-known of these is the so-called *Sexual System*, which was put forward by Linnaeus in 1735 and became almost universally adopted. In this scheme plants were placed into 24 classes, based largely on the number of stamens in each flower, but also utilizing other simple features of the stamens such as their degree of cohesion or relative lengths. Each class was divided into a number of orders, based mainly upon the number of pistils in each flower. Undoubtedly the simplicity of this system, based primarily on a single character, was the major reason for its success, for virtually anyone, with the minimum of training, could allocate a plant to its class and order. The value of the Sexual System does, however, end there. In the class Diandria (two stamens), for example, may be found (amongst others) the genera *Circaea*, *Salvia* and *Anthoxanthum*. Nowadays these genera are placed in three quite different families (Onagraceae, Lamiaceae and Poaceae respectively), the remainder of each of which appeared in Linnaeus' scheme in separate classes (Octandria, Didynamia and Triandria). In other words, a plant belonging to the Diandria had, by definition, two stamens; but nothing else is implied or can be deduced about the plant. It might be an enchanter's nightshade, a sage or a grass. On the other hand, if one discovers that a plant belongs to the grass family (Poaceae), one can immediately infer a great deal of information about it—its peculiar floral morphology and anatomy, its unique seed structure and many other general anatomical, cytological and chemical characteristics. These inferences are likely to prove well-founded, even in respect of a grass not previously investigated, because virtually all grasses possess a basic uniformity which applies equally to undiscovered and to fully studied species. Thus the very statement that a plant belongs to the Poaceae, if correct, automatically predicts many features that will be applicable to that plant. A classification which enables one to carry out this operation successfully is a *predictive classification*, and *predictivity* is one of the most obvious criteria that can be used to assess how good a classification is. Clearly, an artificial system of classification has an extremely low predictivity. Classifications which have high levels of predictivity are termed *natural classifications*, and the groups which they delimit are called *natural groups*. Artificial and natural classifications may be considered opposite extremes in a spectrum of systems in which there are all degrees of intermediates; the more predictive the system the more natural it is considered, the less predictive the more artificial.

In general, artificial systems are based upon fewer characters than natural systems—the most artificial ones are based on a single character and the most natural on a great many. Realization of this has led, in relatively recent years, to attempts by botanists to increase the amount and breadth of data on which their classifications are based, particularly by exploring fields such as cytogenetics, microanatomy and phytochemistry. Such a *multi-variate* approach has not only led to the discovery of a great many new important taxonomic characters, but provides the data upon which numerical taxonomy is based, for this discipline relies upon the provision of a large number of characters (see Chapter 2). In addition, numerical taxonomy usually treats all characters as of equal significance—it does not weight characters as being of lesser or greater taxonomic value. It is possible, however, that some characters are in fact of greater importance than others, and should therefore be weighted positively. Nowadays, there is a great deal of argument (see Chapters 2 and 8) as to whether or not characters should be weighted and, if they should, whether such weighting as may be considered necessary should be done *a posteriori* (retrospectively, in the light of experience) or *a priori* (from the start, according to basic assumption or deduction). Of course, when few characters are used there is always a high degree of weighting, since most characters are given zero-weighting.

General and special purpose classifications

The utilization of a great number of characters follows logically from the observation that more natural or predictive classifications are based on more characters. Thus one might hopefully pose the question: are the best classifications those utilizing the most characters, and hence those which are most predictive? Generally speaking the answer is 'yes'. Such classifications, because they are best for most purposes, are termed *general purpose classifications*, and they are reasonably represented by the familiar sequences of families, genera and species, etc., which appear in Floras. These have arisen over the past two centuries by a process of gradual improvement as more taxonomically valuable characters are discovered and incorporated. Theoretically, this process is never completed, but it obeys the law of diminishing returns, so that the broad outlines of the scheme adopted nowadays, at least that of higher plants, are not likely to be changed.

There are situations, however, when a general purpose classification is not necessarily the best, i.e. when predictivity is not the best criterion. In particular, many botanists wish to know something of the evolutionary history and relationships of a group of taxa; in other words, they will want the classification primarily to reflect the *phylogeny* (phylogenetic or evolutionary pathways) of the plants. A classification which does this is known as a *phylogenetic* or *phyletic classification*, to distinguish it from a *phenetic classification* which is based upon the overall present-day resemblances and differences (not solely structural, but also cytological, chemical, etc.) of the plants.

Theoretically, phylogenetic classifications can be considered less natural than phenetic ones, because they are usually constructed from fewer characters, and features that are thought to be of phylogenetic significance, or lend themselves better to phylogenetic interpretation, are weighted against others. This criticism certainly holds for some of the earlier allegedly phylogenetic classifications, but much less so for more modern ones, including some employing the philosophy and methodology of cladistics (see Chapter 2). Nowadays it should not be so much a case of arguing whether phenetic or phylogenetic classifications are the more predictive or natural, as one of discovering what differences there are between classifications erected according to the two standpoints, and investigating the reasons for and consequences of any differences found. In this way we should be able to discover whether it is possible to construct a classification that is both faithfully phylogenetic and maximally predictive. Despite numerous dogmatic claims to the contrary, we do not yet know the answer to this question.

There are many sorts of special purpose classifications that are certainly very low in predictivity, and which we should therefore describe as artificial. Linnaeus' Sexual System is a special purpose classification and at the time that it was most frequently used it was the best available, as it was easy to understand and brought order to a very confusing subject. Similarly a classification based on flower colour might be the best one to use in a book used by laymen to identify wild flowers, or one based on an alphabetical sequence in a seed catalogue. Thus it may be seen that the merit of a classification is not an absolute criterion, but must be judged according to its purpose. For general purposes predictivity is the criterion, but for other (special) purposes there are other criteria.

Among the most well-known special purpose classifications are those based upon the degree of interfertility between plants; they are of particular interest to biosystematists, geneticists and plant breeders, etc. Two separate schemes have been devised, by Turesson[440] and Danser.[94] Although the terms used are different in each case, and the categories recognized have different definitions, the two schemes have in common the use of categories which are quite different from the family, genus, species, etc., and which are based solely on the extent of interbreeding possible.

The seven terms adopted in the two schemes are defined in Table 1.3. In Turesson's system, the original version of which was modified by Turesson and, later, by others, there are three ranks (*coenospecies, ecospecies, ecotype*) which successively show a greater ability to exchange genes with other taxa of the same rank. The same is true of Danser's system, but the definitions of the terms are such that none of the six is synonymous with any other—Danser's lowest term *convivium* covers the definition of both ecospecies and ecotype, while his higher terms *comparium* and *commiscuum* define the higher and lower aspects of the coenospecies. In addition, Turesson employed a lower term still, the *ecophene*, which denotes an ecological variant, purely the product of environmental modification of the phenotype. Nowadays the general term *habitat modification* or F. E. Clements' term *ecad* are more often used for such variations.

Naturally, it is extremely useful for genecologists and biosystematists to be able to talk of the units which they define by such terms as coenospecies and ecospecies, especially as there is no precise coincidence of these with the

Table 1.3 Comparison of the biosystematic categories of Turesson,[440] Danser,[94] and Gilmour, Gregor and Heslop-Harrison.[156,157]

Behaviour of individuals within the group	Turesson	Danser	Gilmour, Gregor and Heslop-Harrison	Behaviour of group towards other such groups
Individuals capable of hybridizing with one another	Coenospecies	Comparium	Syngamodeme	Group incapable of hybridizing with any other groups
Individuals capable of hybridizing with one another to give hybrids showing some fertility		Commiscuum	Coenogamodeme	Group capable of hybridizing with other groups but such hybrids are sterile
Individuals capable of hybridizing with one another to give fully fertile hybrids	Ecospecies	Convivium	Hologamodeme	Group capable of hybridizing with other groups to give hybrids showing some fertility
Individuals occupying a particular habitat and forming an interbreeding population which differs genotypically from other such populations	Ecotype		Genoecodeme	Group capable of hybridizing with other groups to give hybrids showing complete fertility
Individuals occupying a particular habitat and adapted to it phenotypically but not genotypically	Ecophene		Plastoecodeme	*As previous category*

ranks of the formal taxonomic hierarchy as defined phenetically, or with their phylogenetic pathways. It is important to remember this latter point, and to realize that attempts to redefine the taxonomic species solely on breeding criteria (as is often suggested) is the result of confusion between categories recognized in a general purpose classification with those designed for a special purpose (genecological) classification. In practice, Turesson's terms (especially the ecotype) are quite widely used, but Danser's are rarely encountered. (Care must be taken to avoid the common error of equating the ecotype with a habitat modification.) None of what has been said above should be taken to imply that genecological data are not of use in forming general purpose classifications; indeed, they are usually of some use and frequently of great significance (see Chapters 6 and 7).

Other sorts of special purpose classification are similarly associated with particular fields of study, often far from taxonomic in outlook. The timber trade, for example, employs systems based solely on timber characteristics. As with the biosystematic categories, the timber classifications sometimes coincide with the general purpose classification, but often they do not; the distinction between softwoods and hardwoods is simply that between gymnosperms and angiosperms, whereas the species bearing the various patterns of xylem–parenchyma distribution (apotracheal, paratracheal, etc.) are often disposed in very different botanical families, and so on. Gardeners classify plants according to their life-duration and frost susceptibility (hardy annuals, etc.), ecologists may classify them according to their growth-forms (hemicryptophyte, etc.), and phytosociologists according to their alliances and associations (Mesobromion, etc.).

In all these cases, and others, the data incorporated in the special purpose classifications may well be of value in constructing the general purpose classification, but none should be claimed *a priori* as having special significance.

The point should be made here that the term **taxonomic relationship**, often used very loosely, has a precise meaning only insofar as it is applicable to particular systems of classification; two taxa may be closely related according to one system, but distantly so in another. One can qualify the term by stating that taxa are phylogenetically, or morphologically, or biochemically closely related, etc.

Deme terminology

Gilmour and Gregor[156] proposed a new system of terminology designed to provide an infinitely flexible series of categories which could be used to define any group of individuals on the basis of any set of criteria. This system, the deme terminology, is, in its original concept, non-hierarchical and it falls outside the scope of the formal taxonomic categories (genus, species, etc.). For that reason it avoids the use of root-words such as 'species' and 'type' which are associated with the latter. Central to the idea is the use of a neutral root, a *deme*, which implies nothing in itself except a group of related individuals of a particular taxon. The precise meanings of the terms are provided by various prefixes, of which only three were originally proposed:

topodeme   a deme occurring within a specified geographical area
ecodeme   a deme occurring within a specified kind of habitat
gamodeme   a deme composed of individuals which interbreed in nature.

Later, Gilmour and Heslop-Harrison[157] expanded these suggestions and provided seven further basic terms:

phenodeme   a deme differing from others phenotypically
plastodeme   a deme differing from others phenotypically but not genotypically
genodeme   a deme differing from others genotypically
autodeme   a deme composed of predominantly self-fertilizing (autogamous) individuals
endodeme   a deme composed of predominantly closely inbreeding (endogamous) but dioecious individuals
agamodeme   a deme composed of predominantly apomictic (non-sexually reproducing) individuals
clinodeme   a deme which together with other such demes forms a gradual variational trend over a given area.

Some of these terms (e.g. topodeme, ecodeme, gamodeme) became frequently used, whereas others (e.g. endodeme, clinodeme) are very rarely encountered. A further term, *cytodeme*, has subsequently been used by many workers to indicate a deme composed of individuals all with the same karyotype (chromosome morphology). It is, in fact, one of the main features of this system that further prefixes can be utilized to cover any future possible usage, and there are numerous such examples. In addition, the use of second-order terms was proposed by Gilmour and Heslop-Harrison.[157] Examples of these are *genoecodeme*, an ecodeme differing from others genotypically; and *plastoecodeme*, an ecodeme differing from others phenotypically but not genotypically. The last two terms are the equivalent of Turesson's ecotype and ecophene respectively, as shown in Table 1.3. Also given in Table 1.3 are three other second-order terms, suggested by Gilmour and Heslop-Harrison to cover the higher orders in the genecological classification, equivalent to the comparium, commiscuum and ecospecies. As well as any intrinsic use which they might have, the second-order terms are interesting in demonstrating that the deme terminology may quite easily be adapted into a hierarchical scheme; the possibilities are clearly endless.

In 1963, Davis and Heywood[98] said of the deme system 'It has to be admitted that this basic terminology has so far received only limited support'. This is still true,[52] and it is now clear that it will never become widely adopted.[53] Moreover, the system has been misused by a number of students of animal microevolution, who have called the 'breeding' populations a deme. This is quite contrary to the intention of the original authors, who stated explicitly that it is essential to keep the suffix deme completely neutral—the term gamodeme should be used for deme in the above zoologists' sense. Despite its demise, the deme terminology is of much value in demonstrating the general concept of taxonomic categories, particularly in relation to general and special purpose classification.

Alpha and omega taxonomy

Since modern classifications are ideally based on a very wide range of characters, incorporation of which is a gradual, continuous process, it follows that the naturalness or predictivity of a classification depends on the extent to which the plants have been investigated. Our level of knowledge of plants varies, of course, in different ways. For instance we know far more about the vascular plants than about the lower green plants, about plants of the North Temperate regions than about those of the Tropics, and about plants of great economic value than about those of little commercial interest. Thus our classification of Amazonian liverworts is at a far less advanced (natural) stage than that of cereal grasses or of forage legumes, because it is based on far fewer data.

Valentine and Löve[449] recognized three stages of floristic or taxonomic study: the *exploratory phase*, involving collection and subsequent classification from a limited range of herbarium specimens; the *systematic phase*, when extensive herbarium and field studies of a wide selection of material of each taxon are carried out; and the *biosystematic phase*, during which detailed genetical and cytological studies are made. Davis and Heywood[98] rightly added a fourth stage, the *encyclopaedic phase*, in which data from a very wide range of disciplines are assembled to form a good predictive, natural classification.

Earlier, Turrill[441] had expressed the same idea differently, and perhaps more usefully, because his notion emphasized the continuousness of these phases. Turrill spoke of an *alpha-taxonomy*, equivalent to the first and second of the above four phases, based solely upon more or less obvious external morphological (exomorphic) characters, and an *omega-taxonomy*, which would be the equivalent of an ultimate, perfected system, based upon all available characters. The latter is, almost by definition, unattainable, but it is the distant goal at which taxonomists should aim. Turrill,[442] in 1938, commenting upon attempts in this direction, said 'Some of us please ourselves by thinking we are now groping in a "beta" taxonomy'; in 1963 Davis and Heywood[98] were more certain: 'Today, classification extends beyond "alpha" taxonomy'. In 1987 it would be reasonable to claim that in a few well-studied groups an omega-taxonomy is within reach, but at the same time it should be realized that for the great majority of taxa (particularly lower plants and tropical plants) the 'alpha' stage has not been passed.

The concept of alpha-omega taxonomy ties in well with the view put forward at the beginning of this book that taxonomy is both the most basic and the ultimate field of biology—alpha-taxonomy forms the basis of biology, while the final accumulation of all data is ultimately incorporated into omega-taxonomy. This development of taxonomy, along with a summary of the arguments for and against the different modern approaches, is outlined in the next chapter.

2
The development of plant taxonomy

As mentioned before, this outline of the growth of taxonomy is given only to enable modern taxonomy to be understood more fully, and to be seen in its proper setting. Far fuller accounts are given in other textbooks.[72,255,335] Almost all such accounts attempt to divide historical taxonomy into a number of periods, each marked by a characteristic common goal or principle and frequently separated from one another by notable events, works or authors, such as Linnaeus' *Species Plantarum* (1753), Darwin's *Origin of Species by Means of Natural Selection* (1859), the rediscovery of Mendelian Genetics (1900), or the development of numerical taxonomy (1957). Whilst all these, and others, are notable milestones, none of them really signalled the end of one period and the start of a new one, for taxonomic development has, on the whole, been rather gradual. In any era dominated by one particular outlook there have always been other activities carried on in parallel, or as left-overs from a previous era. Some of the latter, considered very unfashionable at the time, have since come to be regarded in a different light. Furthermore, modern taxonomy has arisen from a number of diverse origins which are not easily treated as a linear sequence.

The phases in the development of taxonomy which are, for convenience, recognized in this account should be viewed with the above considerations in mind.

Phase 1. Ancient classifications

Ancient man must have had systems of plant classification, for he needed to convey to others the names and properties of plants which were of significance to him. Few of these were recorded for posterity and the nearest we can get to understanding them is in the study of the so-called *folk-taxonomies*—classifications which grow up in communities, both primitive and civilized, through need and without the influence of science.[32,33] The degree of detail in these folk-taxonomies is largely marked by the existence or non-existence of vernacular names for taxa, which can also be used to measure the coincidence of the folk-taxa with those of modern taxonomy. In English, the terms grass and sedge, for instance, are largely equivalent to the families Poaceae and Cyperaceae respectively, yet there is no common name for the Ranunculaceae (buttercup, columbine, monk's-hood, etc.). Virtually the only genuine (i.e. non-contrived) common name for the large genus *Carex* is 'sedge', whereas in some groups of much greater human significance the folk-taxonomies distinguish groups far lower even than the species (e.g.

cabbage, sprouts, cauliflower, broccoli, etc., for *Brassica oleracea*; beet, chard, mangold, etc., for *Beta vulgaris*). In other languages similar diversity can be found for plants such as dates, vines and rice.

The first to write down a classification in a permanent and logical form was 'the father of botany' **Theophrastus** (*c*. 370–285 BC), a student of Plato and Aristotle, who became head of the Lyceum in Athens. Theophrastus classified only about 480 taxa, using primarily the most obvious characters of gross morphology (trees, shrubs, subshrubs, herbs) and successively more and more cryptic features. He even recognized differences based on superior and inferior ovaries, fused and separate petals, gross internal anatomy, fruit-types, and so on. Several of the names which Theophrastus used in his *De Historia Plantarum* were later adopted by Linnaeus in his *Genera Plantarum* and are thus still used in the same sense today.

Dioscorides (1st century AD), also Greek, was a physician in the Roman army and therefore interested in the medicinal properties of plants. His *De Materia Medica* described about 600 taxa, largely from first-hand observation, and detailed their useful applications. His book was arranged in a less orderly manner than that of Theophrastus, but nevertheless became the standard reference work for over a millennium and was not completely superseded until the sixteenth century. It can be considered the first herbal.

Phase 2. The herbalists

Through the Middle Ages new written works on botany were very rare and largely based on those of the ancient Greeks. During the Renaissance originality became more of a virtue and the invention of printing in Europe enabled new books to be produced in large numbers. The first of these in the field of botany were the *herbals*, for in those days botany was virtually synonymous with herbalism, the study of plants in relation to their value to man, particularly as foods and medicines, and *herbalists* dominated the sixteenth century botanical world. An excellent general account of herbals is given by Arber,[12] and an extremely thorough treatise on British and Irish herbals and herbalists by Henrey.[195] One of the main features of the herbals was the provision not only of an original text but also of original illustrations (woodcuts) drawn from nature (Fig. 2.1). Some of these herbals provided the rudiments of a natural classification, but this was not their objective and many were arranged wholly artificially, often alphabetically.

Among the most important herbals are those of **O. Brunfels** in 1530, **J. Bock** in 1539, **L. Fuchs** in 1542, **P. Mattioli** in 1544, **W. Turner** in 1551, **M. de L'Obel** in 1570, **J. Gerard** in 1597 and **C. L'Ecluse** in 1601. These were much copied by lesser authors and many such herbals are conspicuously lacking in originality. They also incorporate a great deal of myth and superstition, for they date from a period when plants were supposed to be provided for man by the Almighty—it was man's task merely to discover and utilize their properties. Nevertheless herbals remained popular well after the sixteenth century, for they marked an important stage of development not only in botany and plant taxonomy but also in medicine and pharmacognosy. It is interesting to note that in recent years, during which there has been an

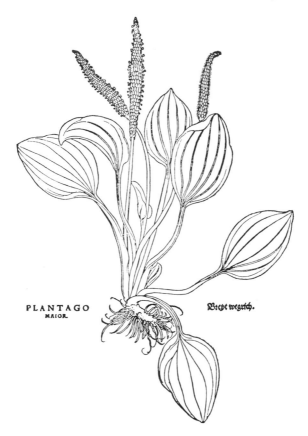

Fig. 2.1 Woodcut illustration of *Plantago major* taken from Fuchs' *De Historia Stirpium* (1542), as reproduced by Arber.

increased awareness of the environment and a desire for 'natural foods', a number of the ancient herbals have been reprinted in popular form along with the appearance of numerous new works, such as R. Mabey's *Food for Free* (1972) and M. Grieve's *A Modern Herbal* (1967), which can justly be considered the modern equivalent of the popular herbal.

Phase 3. The early taxonomists

Towards the seventeenth century plants began to be the focus of attention of a number of scientists for their intrinsic interest rather than for their nutritive or medicinal value. The books which these botanists produced marked an important step forward in plant classification.

A. Caesalpino (1519–1603), an Italian, has been called the first plant taxonomist. His book, *De Plantis* (1583), classified about 1500 species mainly on the basis of growth-habit and fruit and seed form, but it also utilized a whole series of floral and vegetative characters. Thus his arrangement resembled the crude classifications of Theophrastus rather than the artificial system of the herbalists. Caesalpino, using *a priori* or **deductive** reasoning, gave prominence to characters of functional significance, especially those concerned with reproduction but, even so, his book delimited groups of plants closely corresponding to many that we still recognize, for example Brassicaceae and Asteraceae.

The brothers **J. Bauhin** (1541–1631) and **G. Bauhin** (1560–1624) were Swiss botanists who worked separately but along rather similar lines. The most important book which they produced was the latter's *Pinax Theatri Botanici* (1623). Its title (*Pinax* = register) indicates one of its most significant features—the listing not only of all the 6 000 or so species known to him but also all their synonyms, i.e. the various names given to each species by previous workers (Fig. 2.2). Thus the chaotic state of nomenclature which existed was to a considerable degree brought to order. In addition G. Bauhin is notable for his recognition of genera as well as species as major taxonomic levels, and for using a binary nomenclature composed of the generic name followed by a single specific epithet to designate many of the species. Thus, in two major ways, Bauhin's *Pinax* foreshadowed Linnaeus' great works.

❖❖❖❖❖❖❖ ✿ ❖❖❖❖❖❖❖❖❖❖❖❖❖❖❖❖✿❖❖❖✿❖❖❖❖❖❖❖❖❖❖❖❖ ❖❖❖❖ ✿

S E C T I O   S E C U N D A.

HELIOTROPIUM; AURICULA MURIS; ECHIUM; ONO-
SMA; ANCHUSA; LYCOPSIS; BUGLOSSA ET BORRAGO;
Cynoglossum; Cerinthe; Lithospermum; Symphytum; Pulmona-
ria; Consolida media; Brunella; Bellis.

H E L I O T R O P I U M.

H'Ἀιοθρόπιον Dioscoridi lib.4.cap.193.quod τὰ φύλλα τῆ τε ἡλίε συμπεριερέπιτχι, quod folia cum Sole cir-
cumagat : & snoρνίχε &, quod ramuli cum floribus, Scorpionum caudæ modo inflectantur. Theophrafto
7.h·ft.3.& 9. Solaris herba Gaza vertente. Plinio lib.22.cap.21. Heliotropium quod cum Sole fe circum-
agat, etiam nubilo die, tantus fideris amor eft, noctu velut defiderio contrahitur caruleus flos.
Genera duo Dioscoridi : majus flore candido, aut fubfulvo : minus, femine verrucarum (hinc Verrucaria)
penfili. Plinio, tricoccum & Heliostrophium : parvum & magnum intelligens.

I. Heliotropium majus Dioscoridi.	Heliotropium minus,Lob.ico.Ger.repens, Ad.
Heliotropium majus, Matth. Ang. Dod. Gefn.	Heliotropium minus 2 Tab.
hort. (& Scorpioides album) Lac. Adv. Lobel.	Verrucaria altera minor, Lugd.
Cæf.Lugd. (& Verrucaria) Caft.Taber.Camer.	IV. Heliotropium tricoccum.
Ger.Cluf.hift.	Heliotropium minus, Matth.Lac.Caft.Lugd.
Herba cancri majus,Lon.	Dod.Cam.ep.& hort.cui & herba Ciytiæ.
II. Heliotropium minus fupinum.	Heliotropium minus tricoccum , Gefn. hort.
Heliotropium fupinum, Dod.Cluf.hifp.& hift.	Cluf hifp.& hift.
Ger. defc.	Heliotropium vulgare Tornefol Gallorum; fi-
Heliotropium humi fparfum,Lugd.	ve Plinij tricoccon, Ad.
Heliotropium minus 1.Tab.	Helitropium parvum Dioscoridis,Lob.
III. Heliotropium fupinum alterum.	Heliotropium tricoccum,Cæf.Lugd Tab Ger.
Heliotropium minus folio ocimi,Gefn.col.	V. Heliotropium alterum,Matth.Lug.quid?

Fig. 2.2 Part of page 253 of Bauhin's *Pinax Theatri Botanici*, showing the genus *Heliotropium*. The figure is taken from the copy of the 2nd edition (1671) belonging to the Linnean Society of London; this formerly belonged to Linnaeus, whose annotations (mostly his own names for the plants concerned) are seen in the margins. (Reproduced by permission of the Linnean Society of London.)

J. P. de Tournefort (1656–1708) was a Frenchman who carried further Bauhin's promotion of the rank of genus. He had a clear idea of genera, and many of these in his *Institutiones Rei Herbariae* (1700) were later adopted by Linnaeus and are still in use today. Tournefort's system, which classified about 9 000 species into 698 genera and 22 classes, was largely artificial but extremely practical, and it remained 'in force' until superseded by that of Linnaeus, and even later in France where the Linnaean system was never properly adopted.

Nevertheless, his ideas and expertise were in many ways inferior to those of his English contemporary, **J. Ray** (1627–1705), a naturalist who produced several important books on plant classification. Most important of these were *Methodus Plantarum Nova* (1682, 2nd ed. 1703), *Historia Plantarum* (3 vols., 1686, 1688, 1704), and *Synopsis Methodica Stirpium Britannicarum* (1690, 2nd ed. 1696, 3rd ed. 1724 by J. J. Dillenius). The last of these was effectively the first British Flora, and a fine reproduction of it with useful introductory and bibliographic material[416] has been published by the Ray Society, a learned Society concerned with the publication of new and the reprinting of old books on British natural history. The second edition of Ray's *Methodus* dealt with nearly 18 000 species in a complicated system of classification, utilizing a very large number of characters of the flower and vegetative parts, for Ray believed, as we do today, that all parts of the plant should be used in taxonomy. Ray did not develop the idea of binary nomenclature commenced by Bauhin; his species were characterized by phrase-names. His system was somewhat unmanageable, largely due to the great number of species which it attempted to deal with, but technically it was far superior to that of Linnaeus which was to follow 50 years later.

The great increase in the number of species in the books of Ray and his contemporaries was partly due to the recognition of smaller entities, particularly among the lower plants (Ray, for instance, had a narrower species concept than Linnaeus), but largely the result of exploration in regions beyond Europe, so that annually a great many new species were reaching the European taxonomists. It is against this background of increasing diversity of flora and somewhat chaotic bibliographic documentation that Linnaeus' orderly writings must be viewed.

Phase 4. Linnaeus and his apostles

More has been written about the Swede, **Carl Linné** (usually latinized as **Carolus Linnaeus** (1707–1778), than any other biologist except Charles Darwin, which may be taken as a measure of the impact of these two scientists on biology. Linnaeus was the founder of modern taxonomy, both of plants and of animals, and the system of nomenclature which we employ today is essentially his. No attempt is made here to sketch even the outlines of Linnaeus' life, philosophy or writings. Good insight into these (and leads to many of the works on the subject) can be gained by reference to three publications: Stearn's introduction to the Ray Society's facsimile of the first edition of Linnaeus' *Species Plantarum*;[415] Stafleu's *Linnǣus and the Linnaeans*,[413] and Blunt's *The Compleat Naturalist; a Life of Carl Linnaeus*.[40]

As mentioned above, Linnaeus' main contribution was to bring order to the bewildering array of literature, systems of classification and plants which confronted the eighteenth century botanist. He was a prodigious writer and produced many works of great value, which amply illustrate his talent for accurate observation, methodical recording, concise and clear summarizing and energetic enthusiasm.

For plant taxonomists the two most important works are *Genera Plantarum* (1737, with later editions) and *Species Plantarum* (1753, with later editions). In both of these works Linnaeus classified plants according to his artificial 'Sexual System' (Fig. 2.3), an outline of which is given in Chapter 1. This system was first published in 1735 in *Systema Naturae*, a work which classified all known animals and minerals as well as plants. The *Genera Plantarum* listed and briefly described the plant genera recognized by Linnaeus, and hence carried forward the work of Bauhin and Tournefort in giving prominence to the rank of genus. Indeed, many of Linnaeus' generic names came from the works of these two authors, while the new ones were often descriptives or were taken from classical literature or commemorated the names of distinguished botanists, for example *Theophrasta*, *Dioscorea*, *Fuchsia*, *Lobelia*, *Bauhinia*, *Rajania* and even *Linnaea*. About ten editions of *Genera Plantarum* appeared, not all of which were prepared by Linnaeus (several appeared after his death). Taxonomically the most important is the fifth edition (1754), which was very much the work of Linnaeus and was seen by him (and is recognized still by the *International Code of Botanical Nomenclature*) as an adjunct to the two volumes of the first edition of his *Species Plantarum* (1753). Together these two works, both published in Stockholm, cover about 7 700 species in 1 105 genera.

The *Species Plantarum* does not provide generic descriptions (cf. *Genera Plantarum*), but under each genus it numbers, names and briefly describes the various species (Fig. 2.4). References to important earlier literature which also recognized the species are given, together with the synonyms (where applicable), habitats and countries of origin. The description of each species is in the form of a ***phrase-name*** of up to twelve words, commencing with the generic name. The phrase-names were not all new; many were taken from previous works of Linnaeus and others. For new species (i.e. with no synonyms or earlier references), and some little-known others, the entry is concluded by a longer descriptive paragraph; genera with only one species usually were given no description at all for that species, since the generic description (in *Genera Plantarum*, 5th ed.) could equally do for the species.

In addition to the phrase-name, each species was provided (in the margin) with a one-word ***trivial name*** or ***specific epithet***. As demonstrated by several researchers, these names 'originated as an indexer's paper-saving device',[415] but their use proved of such convenience that they quite quickly became the standard name of each species. The specific epithet variously did or did not repeat a key word from the phrase-name. Such a binary name, formed of the generic name plus the specific epithet, is known as a ***binomial***, as distinct from the phrase-name which is a ***polynomial***. Binomials were quite often used by G. Bauhin in his *Pinax* (1623) over a century earlier, so it is not true to say that Linnaeus devised the system. However, it was their appearance as an additional, convenient name for every species in *Species Plantarum* which led to their eventual universal adoption, and nowadays they are mandatory.

REGNUM VEGETABILE
CLAVIS SYSTEMATIS SEXUALIS
NUPTIAE PLANTARUM.
Actus generationis incolarum Regni vegetabilis.
Florescentia.
⌠*PUBLICÆ.*
Nuptiae, omnibus manifestae, aperte celebrantur.
Flores unicuique visibiles.
⌠Monoclinia.
Mariti & uxores uno eodemque thalamo gaudent.
Flores omnes hermaphroditi sunt, & stamina cum pistillis in eodem flore.
⌠Diffinitas.
Mariti inter se non cognati.
Stamina nulla sua parte connata inter se sunt.
⌠Indifferentismus.
Mariti nullam subordinationem inter se invicem servant.
Stamina nullam determinatam proportionem longitudinis inter se invicem habent.

1. MONANDRIA.	7. HEPTANDRIA.
2. DIANDRIA.	8. OCTANDRIA.
3. TRIANDRIA.	9. ENNEANDRIA.
4. TETRANDRIA.	10. DECANDRIA.
5. PENTANDRIA.	11. DODECANDRIA.
6. HEXANDRIA.	12. ICOSANDRIA.
	13. POLYANDRIA.

⌊Subordinatio.
Mariti certi reliquis praeferuntur.
Stamina duo semper reliquis breviora sunt.
14. DIDYNAMIA.
⌊Affinitas.
Mariti propinqui & cognati sunt.
Stamina cohaerent inter se invicem aliqua sua parte vel cum pistillo.

16. MONADELPHIA.	19. SYNGENESIA.
17. DIADELPHIA.	20. GYNANDRIA.
18. POLYADELPHIA.	

⌊Diclinia (a δίς bis & κλίνη thalamus s. duplex thalamus).
Mariti & Feminae distinctis thalamis gaudent.
Flores masculi & feminei in eadem specie.

21. MONOECIA.	23. POLYGAMIA.
22. DIOECIA.	

⌊Clandestinae.
Nuptiae clam instituuntur.
Flores oculis nostris nudis vix conspiciuntur.
24. CRYPTOGAMIA.

Fig. 2.3 Linnaeus' 'Sexual System' of plant classification, copied from the 10th edition of *Systema Naturae* (1759).

PENTANDRIA

MONOGYNIA.

HELIOTROPIUM.

indicum.   1. HELIOTRPIUM foliis cordato-ovatis acutis fcabri-
ufculis, fpicis folitariis, fructibus bifidis. *Fl. zeyl.* 70.
Heliotropium foliis ovatis acutis, fpicis folitariis. *Hort.
cliff.* 45. *Roy. lugdb.* 405.
Heliotropium americanum cæruleum. *Dod. mem.* 83.
Pluk. phyt. 245. *f.* 4.
β. Heliotropium americanum cæruleum, foliis hormini an-
guftioribus. *Herm. lugdb.* 307. *Sloan. jam.* 98.
Habitat in India *utraque.* ☉

europæum.   2. HELIOTROPIUM foliis ovatis integerrimis tomen-
tofis rugofis, fpicis conjugatis. *Hort. upf.* 33. *Sauv.
monfp.* 305.
Heliotropium foliis ovatis integerrimis, fpicis conjunctis.
Hort. cliff. 45. *Roy. lugdb.* 404.
Heliotropium majus diofcoridis. *Bauh. pin.* 253.
Habitat in Europa *auftrali.* ☉

fupinum.   3. HELIOTROPIUM foliis ovatis integerrimis tomen-
tofis plicatis, fpicis folitariis,
Heliotropium minus fupinum. *Bauh. pin.* 253.
Heliotropium fupinum. *Cluf. hift.* 2. *p.* 47.
Habitat Salmanticæ *juxta agros*, Monfpelii *in litto-
re.* ☉

Fig. 2.4   Part of page 130 of Linnaeus' *Species Plantarum* (1753), showing the start of Class V, Pentandria, and of the genus *Heliotropium*. Note the references to Bauhin (see Fig. 2.2) under the second and third species.

Since *Species Plantarum* (1753) was the first work in which all plants were given a Latin binomial, this publication is today taken as the starting point for the nomenclature of all green plants, except mosses and a few special groups of algae (see Chapter 9).

Thus in 1753 the botanical world was provided for the first time with a work which catalogued all known plants, provided them all with a binomial as well as a polynomial, listed many of their known earlier synonyms, and disposed them in a simple, logical (albeit artificial) hierarchical system of classification. Later (1762) Linnaeus issued a second edition of *Species Plantarum*, which not only corrected most of the numerous minor mistakes, etc. of the first edition, but also revised taxonomic opinions and added numerous newly discovered species. In 1764 he also produced the 6th edition of *Genera Plantarum*, which is to be regarded as associated with the second edition of *Species Plantarum* just as the 5th edition was with the first edition of the latter. The third edition of *Species Plantarum* (1764) scarcely differed from the second, but in 1767 Linnaeus published a *Mantissa* ('Supplement') *Plantarum* and in 1771 a *Mantissa Plantarum Altera*, which, together with his son's (Carl Junior) *Supplementum* of 1781 (partly consisting of species described by Carl Senior), form a supplement to the 2nd and 3rd editions of *Species Plantarum*.

The adoption of Linnaeus' system (both classification and nomenclature) in countries outside Holland (where Linnaeus obtained his degree and spent many years working and where many of his books were published) and Sweden varied a good deal. Some of the earliest works following that system used its classification and polynomials, but not its binomials. An important milestone in Britain was the appearance of W. Hudson's *Flora Anglica* (1762), a work succeeding Dillenius' third edition of Ray's *Synopsis* (1724) as the standard British Flora, and the first to adopt entirely the Linnaean system (including binomials). In France, Linnaeus' Sexual System was never taken up by the majority of the leading taxonomists, although his binomial nomenclature was adopted more or less at the same time as it was in England. Linnaeus' method was promulgated not only by Linnaeus' own publications, but also by those of his many correspondents and students ('apostles', as they are called by Stafleu[413]). Among these may be named **W. Curtis, J. Banks** and **J. E. Smith** in England and **J. F. Ehrhard, P. C. Fabricius, P. Forsskål, P. Kalm, P. Loefling, D. C. Solander** and **C. P. Thunberg** elsewhere, several of whom undertook botanical exploration in various distant parts of the world.

After the death of Linnaeus' son in 1783 the father's collections, including books, were purchased by J. E. Smith, and in the following year they were shipped to England. Smith founded the Linnean Society of London in 1788 and upon his death in 1828 the collections were sold to the Society. Today the 'Linn. Soc.' is a leading international society concerned with the fostering of the science of natural history, particularly taxonomy, and still houses the Linnaean collections, the object of visits by taxonomists from all countries of the world, in Burlington House, Piccadilly, London.

Linnaean classification continued to dominate taxonomic works until well into the nineteenth century. In Britain, for instance, the great majority of Floras up until about 1825 employed the Sexual System, for example J. E. Smith's *English Botany* (1790–1814), often known as Sowerby's *English Botany*, since the latter drew the plates. (The best known version of this *Flora* is the third edition, 1863–1886, text by J. T. B. Syme, which does not, of course, use the Sexual System). A fourth edition of *Species Plantarum* was prepared in Berlin, mainly by **C. L. Willdenow,** and published between 1797 and 1830. Its six volumes (in 13 parts) covered over 16 000 species, more than twice as many as in the first edition, and they represent the last successfully completed world monograph not only of all plants together, but also of such important monocotyledonous families as the Poaceae.

Phase 5. Post-Linnaean natural systems

Linnaeus' Sexual System was popular largely because of its simplicity. In fact it did not classify plants in quite the most slavishly simple manner that could be envisaged (i.e. only on the number of stamens and pistils), because the classes 14–23 were based on somewhat more subtle characters than stamen number alone. Hence *Digitalis*, the foxglove, which has four stamens, appeared not in the Tetrandria, but in Didynamia, with two long and two short stamens. Thus there is slightly less discrepancy between the groups of the Sexual System and those of modern systems than would otherwise have been the case. Nevertheless, the Sexual System was very artificial. Linnaeus himself was fully aware of this, and sought at various times to present the rudiments of a natural system based on overall similarities, as had Ray and others before him. In 1738 he formulated an outline of a possible system and in 1764 he listed 58 'natural orders' in an appendix to the sixth edition of *Genera Plantarum*. These 'natural orders' are what we today call families, and many of them corresponded closely with modern families.

The foundation of the modern families comes mainly, however, from the work of French taxonomists in the latter part of the eighteenth century, notably **M. Adanson** (1727–1806), **A.-L. de Jussieu** (1748–1836) and **J. de Lamarck** (1744–1829), who never followed the Sexual System.

Adanson produced his *Familles des Plantes* in 1763. Today he is most remembered for championing the idea that in classification one should use a great range of characters covering all aspects of the plant, and without placing greater emphasis on some than on others. This is called an *empirical* approach. He was a severe critic of Linnaeus' works, and considered Tournefort's classification, upon which he improved, far superior. Adanson recognized 58 families of plants, many of them with the same circumscription as today.

A.-L. de Jussieu was a student of his uncle, B. de Jussieu, as was Adanson. In his most significant work, *Genera Plantarum* (1789), he divided plants into three groups: Acotyledons (cryptogams plus a few misunderstood monocotyledons); Monocotyledones (monocotyledons); and Dicotyledones (dicotyledons and gymnosperms). Within the last two groups he used many of the familiar modern characters (superior versus inferior ovaries, stamens free versus attached to corolla, fused versus free petals, etc.), so that plants as a whole were divided into 15 classes and 100 natural orders (families). Although several of these were very artificial, a good proportion are still found similarly delimited in modern classifications, and de Jussieu's *Genera* can be regarded as the immediate progenitor of the modern system.

Lamarck is best known for his theory of evolution, Lamarckism, whereby characters acquired during life become inherited. This theory was, however, finalized after his main taxonomic contributions appeared. These comprised his *Flore Françoise* (1778, with later editions) and the botanical part of Panckoucke's *Encyclopédie Méthodique* (1783–1798, completed by J. L. M. Poiret, 1804–1817). Lamarck's taxonomic fame rests not on his system of classification (the *Encyclopédie*, while arranged alphabetically, advocated the de Jussieu system) but upon his realization that a natural system of classification was not the best for rapid identification. Accordingly, his *Flore Françoise* contained an analytical method, closely similar to the dichotomous keys of modern Floras, which was meant to be simply a method of identification.

The work epitomized by these three Frenchmen, plus progress in an understanding of plant anatomy and physiology, the greatly increased exploration of the tropics and southern hemisphere and the cultivation in European gardens of many of the plants brought back, and the development and widespread use of better optical aids, paved the way for the nineteenth century taxonomists.

Before the nineteenth century knowledge of plant structure was confined to exomorphic aspects and to a very elementary outline of endomorphic features, which had been established by the seventeenth century anatomists such as **N. Grew** and **M. Malpighi** in their works of the 1670s. **R. J. Camerarius** had in 1694 experimentally demonstrated the male and female nature of the stamens and carpels, and **T. Fairchild** in 1717 had produced what was probably the first experimental interspecific hybrid (between a Carnation and a Sweet William, two species of *Dianthus*). Nevertheless, little detail of sex in plants was grasped, and hence the true (and widely divergent) nature of cryptogams was unknown. In the nineteenth century these topics received a great deal more attention in various countries of Europe. The names of three men should be mentioned: **R. Brown**, who in the early part of the century investigated the detailed floral morphology of many groups, recognized the naked ovule of the gymnosperms, and discovered the nature of the cell nucleus; **W. Hofmeister**, who in the mid part of the century worked out details of the life cycle of many cryptogamic groups, especially bryophytes and pteridophytes; and **E. Strasburger**, who later in the century carried further Hofmeister's work and was the first to observe sexual fusion in higher plants.

Exploration of Asia, Africa and America was providing an enormous number of new species. This can be illustrated by reference to the works of **F. H. A. Humboldt**, **A. J. A. Bonpland** and **C. S. Kunth** on the South American tropics. Of their many impressive works, *Nova Genera et Species Plantarum* (1816–1825) in 7 volumes and 36 parts is the greatest, and it led to the production, started by **C. F. P. von Martius**, of *Flora Brasiliensis* (1840–1906). They had carried forward a thirst for scientific exploration so well exhibited by James Cook and the natural historians who accompanied him on his three great voyages (1768–1780). Classifications which appeared during the nineteenth century were therefore devised in a lively and dynamic environment, and not surprisingly there were many of them.

A. P. de Candolle (1778–1841), a Swiss botanist, contributed much to the development of higher plant classification, as well as to other fields of plant science. His book *Théorie Élémentaire de la Botanique* (1813), in which he first introduced the word *taxonomy*, set out an outline classification of plants as well as the principles which he thought should govern classification; a good account of these is given by Cain.[58] He divided plants into two major groups: Cellulares and Vasculares, the non-vascular and vascular plants respectively. His most important work was *Prodromus Systematis Naturalis Regni Vegetabilis* (1823–1873), a 17-volume book written by 35 authors under the general editorship of A. P. de Candolle (Vols. 1–7) and, after the latter's death, by his son, **A. de Candolle** (1806–1893). This covered all species of dicotyledons in the world, accounting for over 58 000 species in 161 families (Fig. 2.5). Included in these were gymnosperms, still placed close to the catkin-bearing dicotyledons, but the work remained unfinished in that the monocotyledons and cryptogams were never covered. This mammoth undertaking represents the most modern world monograph of dicotyledons, and for many individual families no later ones have since appeared. De Candolle also published a great number of family and genus monographs, most of them dating from before his *Prodromus*. De Candolle's system was an improvement upon that of de Jussieu, and in many general respects it resembles many twentieth century schemes.

Most other nineteenth century taxonomists, at least before those of the

532   DICOTYL. seu EXOGENÆ.

XII. HELIOTROPIUM *Tourn. inst.* 138. *t.* 57. *Linn. gen. n.* 179. *Juss. gen.*
130. *Gœrtn. fruct.* 1. *p.* 329. *t.* 68. *f.* 2. *Lam. ill. t.* 91. *Lehm. asp. p.* 19
(*excl. in omnibus spec. nonnull.*) *Spenn. gen. fl. germ. ic. et descr.*

Calyx 5-partitus aut rarissime 5-dentatus persistens. Cor. hypocrateri-
morpha, fauce pervia interdum barbata, limbi laciniis plicaturà simplici
vel rarissime dente interjecto donatis. Stylus brevis. Stigma subconicum.
Nuculæ uniloculares juniores basi cohærentes demum separabiles basi
clausæ. Receptaculum commune nullum. Semina exalbuminosa, em-
bryone inverso, cotyledonibus planis (1). — Herbæ aut suffrutices nunc
varie villosæ rarius glaberrimæ. Folia integra aut denticulata, alterna aut
rarius opposita. Spicæ unilaterales. Cor. albæ aut purpurascentes,
interdum per exsiccationem ochroleucæ nunquam luteæ. — Sectiones
plures (omnes?) forte in genera convertendæ ?

Sectio i. Catimas *Alph. DC.* — Sect. Euheliotropii spec. *DC. mss.*

Nuculæ 4, ovoideo-triangulares, intus nempe angulares, dorso convexæ, la-
teribus non bifoveolatæ. Corollæ faux imberbis ; lobi angusti, æstivatione intra
tubum inflexi. Stigma conico-truncatum, apice hispidum, simplex, vel subbifidum.
Spiculæ bifurcatæ, ebracteatæ, juniores apice scorpioideæ — Nomen ex χατὰ
deorsum, infra, et ἱμας, αυτος lacinula, quia lacinulæ cor. in æstiv. inflexæ.

* *Antheræ prope basim corollæ. Stylus brevissimus, glaber.*

1. H. grandiflorum (Auch.! pl. exs. n. 2362 et 2376, non Schranck), herba-
ceum totum molliter villoso-canescens erectum, foliis petiolatis ovalibus obtusis
integerrimis, spicis solitariis ebracteatis, corollæ tubo pubescente calycis lobis
lanceolato-linearibus duplo longiore, nuculis subrugulosis glabris. (I) in Ar-
meniâ legit cl. Aucher! (2). Radix parva sublignosa. Caulis 6-7 poll. longa. Cor.
alba 3-4 lin. longa, lobis oblongis subacutis. (v. s. a cl. inv.)

Fig. 2.5 Part of volume IX page 532 of de Candolle's *Prodromus* (1845), showing the
start of the genus *Heliotropium*.

next phase, used de Candolle's system or devised modifications or extensions
of it. Among them should be mentioned **J. Lindley** (1799–1865), **A. T.
Brongniart** (1801–1876) and **S. L. Endlicher** 1805–1849). Endlicher's *Genera
Plantarum* (1836–1841) covered 6 835 genera of plants, and separated the
Thallophyta (algae and fungi) from the higher plants (Cormophyta). Whereas
de Candolle's sequence of dicotyledon families had started with the Ranunculaceae, Endlicher's commenced with the apetalous families.

The last major natural classification was that of **G. Bentham** (1800–1884)
and **J. D. Hooker** (1817–1911). Their *Genera Plantarum* (1862–1883) dealt
only with seed-plants, commencing with the Ranunculaceae and related
families such as the Magnoliaceae and dealing successively with dicotyledons,
gymnosperms and monocotyledons. It described 200 families and 7 569
genera, each in meticulous detail and many subdivided into subgenera and/or
sections. The dicotyledons were divided into three great groups: Polypetalae
(with free petals), Gamopetalae (with fused petals), and Monochlamydeae
(with no petals). Bentham was an extremely accomplished self-trained
classical taxonomist who also wrote many important monographs (e.g.
Labiatarum Genera et Species, 1832–1836, in 783 pages) and Floras (e.g.
Handbook of the British Flora, 1858, later editions revised by J. D. Hooker;
Flora Australiensis, 1863–1878, in 7 volumes). Hooker, like his father before

him, was Director of the Royal Botanic Gardens, Kew, and explored many parts of the world and wrote several valuable Floras (e.g. *The Student's Flora of the British Islands*, 1870; *The Flora of British India*, 1872–1897, in 7 volumes). Their *Genera Plantarum*, with a general outline based upon that of de Candolle, set new standards in descriptive botany, having been drawn up afresh from the study of herbarium specimens, and it became the standard work in many parts of the world. Indeed, the great herbaria at the British Museum, Kew and Paris are still arranged according to it, and Bentham and Hooker's *Handbook of the British Flora* remained the major British Flora until 1952.

In contrast, rather little progress in the classification of cryptogams was made during this phase. In the immediate post-Linnaean era *Lycopodium* was classed as a moss, cycads among the ferns, *Salvinia* as a liverwort, and *Equisetum* among the conifers. In 1813 de Candolle was the first to place all the pteridophytes together as a separate group. In 1843 Brongniart advocated the division of plants into two main groups, Cryptogamae and Phanerogamae, while **A. Braun** in 1864 recognized three groups: Bryophyta (algae, fungi, and bryophytes); Cormophyta (pteridophytes); and Anthophyta (spermatophytes). Real progress towards the modern system was not made until Eichler's classification of 1883, which belongs to the next phase. Nevertheless, within each of the main groups of lower plants a greater understanding of the degrees of diversity was arising, mainly from the study of microscopic structures and of the life cycles. Naturally, knowledge of algae (and fungi) lagged well behind that of bryophytes and pteridophytes.

Phase 6. Post-Darwinian phylogenetic systems

The evolutionary theories of **C. Darwin** and **A. R. Wallace**, even the former's *The Origin of Species by Means of Natural Selection* (1859), had little immediate effect on plant classifications. Thus Bentham and Hooker's *Genera Plantarum* was produced well after Darwin's theory was published, yet (despite some misgivings by the junior author) in no way did it aim to be a phylogenetic system. Slowly, however, the idea took hold that, since plants had gradually evolved along diverging and converging pathways, their classification should reflect those pathways. The rediscovery of G. Mendel's laws of inheritance in 1900 and the development of the modern theory of chromosomes at about the same time led to the desire for phylogenetic classifications.

The construction of phylogenetic classifications, however, runs into two great practical problems, neither of which has been satisfactorily solved. Firstly, it is rarely possible to reconstruct with *certainty* the past evolutionary pathways; and, secondly, even if one could it is hardly possible to devise a satisfactory method of designating a branching pattern by means of a single linear sequence, which is needed in a Flora or other systematic treatment, although a number of formats have been proposed.[477] Many attempts have been made to produce such systems, the aim being to construct a sequence starting with the most *primitive* or least specialized and ending with the most *advanced* or derived, and to ensure that each taxon recognized is *monophyletic*, i.e. has arisen by the diversification of a single ancestor (as opposed to

polyphyletic, arising from more than one ancestral group). The concepts of monophyly and polyphyly are, of course, relative, for it is possible that all organisms are monophyletic if one traced back their ancestors far enough. Hence the statement that a taxon is monophyletic usually implies that it is considered that all the constituents of that taxon have a common ancestor which would also be classified within that same taxon. Hennig[193,194] has sought to define the terms monophyly and polyphyly more precisely by the introduction of a third category, paraphyly. ***Paraphyletic*** groups would be considered monophyletic in schemes recognizing only monophyly and polyphyly, but they resemble polyphyletic groups in that both lack a single exclusive ancestor (Fig. 2.6). In other words, a paraphyletic group is one in which all the members possess a single ancestor in common, but which does not constitute *all* the descendants of that ancestor.

The beginning of this sixth phase of activity coincided with a time when our

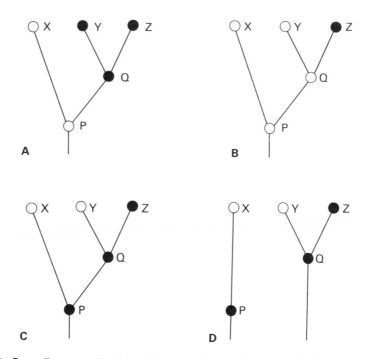

Fig. 2.6 Four diagrams showing different origins of three species (X, Y, Z) from the ancestral taxa P and Q in order to illustrate the concepts of monophyly, paraphyly, polyphyly, parallelism and convergence. The possession of one or other of two contrasting character-states by each of the five taxa is indicated by an open or closed circle respectively. **A.** Groups YZ and XYZ are both monophyletic; the similarity between Y and Z is a synapomorphy; the difference between X and YZ is due to divergence. **B.** Group XY is paraphyletic; group XYZ is monophyletic; the similarity between X and Y is a symplesiomorphy; the difference between Y and Z is due to divergence. **C.** Group XY is polyphyletic; group XYZ is monophyletic; the similarity between X and Y is a false synapomorphy caused by parallelism. **D.** Groups XY and XYZ are both polyphyletic; group YZ is monophyletic; the similarity between X and Y is a false synapomorphy caused by convergence.

knowledge of the life cycles of all the main groups of plants was complete in outline, when the importance of sex and hybridization in plants was reasonably well grasped and when characters other than exomorphic features were being increasingly utilized by taxonomists. For example, the author widely cited as the first to use endomorphic anatomy as a standard taxonomic tool is **E. Bureau** in his *Monographie des Bignoniacées* (1864). Without these developments the construction of a phylogenetic classification would have been impossible.[443]

The earliest phylogenetic classification of plants is usually reckoned to be that of Martius' former assistant, **A. W. Eichler** (1839–1887), first published in detail in the third edition of his *Syllabus der Vorlesungen über Specielle und Medicinisch-Pharmaceutische Botanik* (1883). However, some drawings of 'genealogical trees' had been produced earlier by E. Haeckel (*Monophyletischer Stammbaum der Organismen* and *Stammbaum des Pflanzenreichs*, both published in Jena in 1866). In Eichler's scheme the Cryptogamae and Phanerogamae were still retained as the two major groups of plants, the former being divided into Thallophyta, Bryophyta and Pteridophyta, and the latter into Gymnospermae and Angiospermae. This scheme at once differed significantly from those previously held in highest esteem, mostly derived from de Candolle. In particular the gymnosperms preceded the angiosperms and the monocotyledons preceded the dicotyledons, implying the earlier group in each pair was the more primitive or ancient. The dicotyledons were divided into only two groups, those with free petals (Choripetalae) and those with fused petals (Sympetalae), the old Monochlamydeae (or Apetalae) being dispersed among them, and the catkin-bearing plants were placed at the start of the dicotyledons, suggesting that they were thought to be the most primitive. Nevertheless, it is unlikely that Eichler's sequence, or that of Engler which followed, was actually intended to reflect phylogeny directly; these authors probably attempted a rearrangement of the earlier natural systems to accord more closely with their phylogenetic theories.

Eichler's classification formed the basis of that of the great German systematist **H. G. A. Engler** (1844–1930), who dominated the plant taxonomic world at the turn of the twentieth century. His scheme first appeared in a relatively minor publication of 1892, and was incorporated into his three *magna opera: Die Natürlichen Pflanzenfamilien* (1887–1915, with K. A. E. Prantl; second edition 1924 →, incomplete), which comprised a revision down to the generic (and often subgeneric or sectional) level of all groups of plants; *Syllabus der Pflanzenfamilien* (1st edition 1892, latest complete (twelfth) edition by H. Melchior *et al.*, 1954 and 1964, in two volumes), a revision of all plants down to family (and often subfamily or tribe) level; and *Das Pflanzenreich*, 1900–1953, edited by Engler until his death but written by many authors. *Das Pflanzenreich* was an attempt to provide an up-to-date *Species Plantarum* covering all plants, but it was never completed. Of 208 families of seed-plants originally projected, accounts of 78 actually appeared, as well as a single cryptogamic (moss) family (Sphagnaceae); together these were included in 107 volumes. Nevertheless, for a great number of families, *Das Pflanzenreich* represents the latest world monograph, and for many other families *Die Natürlichen Pflanzenfamilien* is the latest monograph to generic level.

Because of the long period of development and use, Engler's system underwent many changes and, under the guise of recent editions of *Syllabus der Pflanzenfamilien*, it is still changing. One may, however, generalize to some degree. Plants were disposed in a number of divisions; there were 13 divisions in 1919, 14 in 1936 and 17 in 1954. Most of these were algal, as the algae were for the first time separated into a number of distinct divisions. In the earlier versions higher plants were placed in two divisions: Embryophyta Asiphonogama (bryophytes and pteridophytes) and Embryophyta Siphonogama (gymnosperms and angiosperms), but in the latest version these two pairs are both split. As in the Eichler scheme, the monocotyledons preceded the dicotyledons (except in the latest, 1964, version), the Archichlamydeae (= Choripetalae or Polypetalae) preceded the Metachlamydeae (= Sympetalae or Gamopetalae), and the Archichlamydeae commenced with the catkin-bearing plants and their relatives (Amentiferae).

Apart from his work on the above three series, Engler and his many colleagues and students wrote an enormous number of floristic, monographic and phytogeographic volumes of much importance. Engler's system was very widely adopted, and many of the world's great herbaria are arranged by various versions of it, for example Berlin, Arnold Arboretum and Leningrad, as are various modern Floras, for example *Flora Europaea* (1964–1980).

C. E. Bessey (1845–1915) was the first American to make a major contribution to plant classification. His system (*Outlines of Plant Phyla*, 1911) resembled Engler's in the lower groups, but differed in that the pteridophytes and gymnosperms (like the algae) were each divided into a number of divisions, and the dicotyledons were subdivided in a totally different way, commencing with the Magnoliaceae–Ranunculaceae assemblage. In this last respect the Bessey arrangement recalls those of de Candolle and of Bentham and Hooker, but it and Hallier's scheme (q.v.) were the first allegedly phylogenetic systems to start with the Ranunculaceae and/or Magnoliaceae and therefore to implicate them as the most primitive dicotyledons. Bessey, like Eichler and Engler, preceded the latter by the monocotyledons, although nowadays the reverse is almost invariably the case.

H. Hallier (1868–1932) independently produced in 1905 a scheme for the flowering plants which resembled Bessey's in several respects, particularly in commencing the dicotyledons with the Ranunculaceae–Magnoliaceae assemblage, but it differed in that the monocotyledons followed the dicotyledons.

Since the days of the early phylogenetic taxonomists a great number of classifications have been proposed, often differing only in small degree from a previous one and usually with both good and bad features. With respect to the flowering plants, almost all of them are loosely referable to one of the two schemes outlined above, i.e. the Eichler–Engler system or the Hallier–Bessey system, the latter in turn derivable from the de Candolle–Bentham and Hooker systems. Authors responsible for Eichler–Engler-based schemes include **E. Warming** (1895), **R. von Wettstein** (1911), **A. B.Rendle** (1930), **A. A. Pulle** (1938), and **C. Skottsberg** (1940), while Hallier–Bessey-based schemes include those of **J.Hutchinson** (1926–1934), **A. L. Takhtajan** (1943), **C. R. de Soó** (1953), **L. D. Benson** (1957), **A. Cronquist** (1957), **R. F. Thorne** (1968) and **R. Dahlgren** (1975). In fact most of these two groups of taxonomists have more or less confined their attentions to the flowering plants. Their adjustments to existing schemes have been made in the light of new knowledge, which has suggested a different evolutionary pattern from that previously suggested. Adjustments are mostly of two types: a rearrangement of the grouping of families into orders or a change in the sequence of the families, arising from new ideas or evidence on the phylogenetic pathways; and a splitting of families or orders when evidence appears to suggest that a taxon is not monophyletic. An illustration of the latter case is given by the latest (eighth) edition of J. C. Willis's *A Dictionary of the Flowering Plants and Ferns* (revised by H. K. A. Shaw, 1973). This contains many families which have recently been erected (often of one or few genera) on such grounds, and which had formerly been included in scarcely any or no schemes of classification, for example Foetidiaceae, Uapacaceae. Shaw recognized 539 families of seed-plants (gymnosperms and angiosperms).

The use of the various sorts of evidence, and the extent to which new data are utilized in these classifications, have mostly been decided by highly subjective means. The dangers of this can be illustrated by one aspect of Hutchinson's scheme,[218] which differs from all other schemes proposed in that the dicotyledons are divided into two great groups: those characteristically and primitively woody (Lignosae), and those characteristically and primitively herbaceous (Herbaceae). In Hutchinson's sequence the former precede the latter, so that the Magnoliaceae commence the dicotyledons, while the herbaceous Ranunculaceae appear over halfway through. It must be emphasized that this is not a reversion to the pre-Linnaean form-classifications, whereby plants were divided simply according to growth-habit; there are many herbaceous species in Hutchinson's Lignosae and many woody ones in his Herbaceae. Nevertheless, in his insistence, during half a century and three editions of his *The Families of Flowering Plants*, that the dicotyledons diverged at an early stage of their evolution into two main groups, one woody and the other herbaceous, Hutchinson (1884–1972) has produced a system which has no other adherents and which separates pairs of families, for example the Araliaceae from the Apiaceae, and the Verbenaceae from the Lamiaceae, considered by virtually all other taxonomists to be very closely related.

Despite this subjectivity, there has in the past twenty years or so been a convergence in the broad outline of modern phylogenetic classifications (with the exception of that of Hutchinson and a few other eccentric schemes), and even the 1964 version of Engler's system[285] bore a much closer resemblance to the Hallier–Bessey-based schemes than had previously been the case. Of all the above schemes, those of Cronquist[77] (latest version 1981), Takhtajan[436] (latest version 1987), Thorne[438] (latest version 1983), and Dahlgren[92,93] (latest full version 1983) have attracted most attention in recent years; the first two are in fact extremely similar. In all of them the dicotyledons precede the monocotyledons, and the Magnoliaceae or a close relative commence the former.

Hutchinson considered that the resemblance of families such as the Verbenaceae and Lamiaceae is the result of evolutionary *convergence* (Fig. 2.6), the possession of similar characteristics in two or more groups without an immediate common ancestor, presumably arising as a response to evolutionary pressures. For example, self-pollinating species (which do not need to attract insects) from genera (or even families) whose cross-pollinating representatives are totally different may come to resemble one another quite closely, at least superficially, through the loss or reduction of petals, reduction in number and size of stamens, loss of nectaries, scent and bright colours, and the assumption of a relatively dwarf and short-lived habit. A related phenomenon is evolutionary *parallelism* (Fig. 2.6), the possession of similar characteristics by two or more taxa which do have a common ancestor, but in which those characteristics are absent. Examples are the vicariant endemics discussed by Richardson[359] and the polytopic amphidiploids mentioned by Stace,[405] both of which have arisen from the same parents independently and in different places. Many examples of convergence and parallelism are known, but it is not always easy to be sure whether two taxa showing close similarity are the result of either of these phenomena, or are truly closely phylogenetically related. It is the belief of most biologists that convergence and parallelism are manifest in only a small proportion of the total characteristics (albeit often the more conspicuous ones), so that an exhaustive survey of the taxa would demonstrate the true situation. Characteristics likely to be the result of convergence or parallelism (mostly those known to be specializations in respect of particular modes of behaviour or ecological niches) are generally avoided by taxonomists attempting to construct phylogenetic classifications.

One method of inferring phylogeny is to pinpoint primitive (i.e. ancient) as opposed to advanced (i.e. recent) characters and to assign primitiveness to taxa which possess high proportions of the former. This is a justifiable procedure (on the grounds that it represents more or less the only one available *in extenso*) so long as one does not, at the level of the *taxon*, equate primitiveness with ancientness. A taxon possessing many primitive (ancient) characters might in fact be a relatively recent taxon which has retained primitiveness for the most part, but has become specialized in certain directions, or one which had recently attained characters generally indicative of primitiveness. In other words, a character may evolve at very different rates in different taxa, and within one taxon various characters may evolve at very different rates, and in different directions.

Perhaps the first to utilize this concept consciously was Bessey,[34] who forwarded seven general dicta and 21 dicta concerned with the vegetative and floral structure of angiosperms. Hutchinson[217] laid down 24 general principles and several others have made similar attempts. **K. R. Sporne** has sought less subjective means. By simple statistical tests he has compiled a list of characters which are significantly positively correlated in their occurrence among dicotyledon families and also a list of their opposite states which are similarly positively correlated. After some 40 years' work, he accumulated 26 such pairs of characters.[398,400] The decision as to which list represents the primitive and which the advanced state was made primarily by examining the fossil record, to discover which characters are nowadays more, and which less, common than earlier in the fossil record. By use of these data, an *advancement index* was constructed for each family, the score of 0 (most primitive) to 100 (most advanced) being based upon the proportion of advanced characters that a family contains. On this basis, among the dicotyledons, the Magnoliaceae as defined by Cronquist has the lowest advancement index (20), and four families (Callitrichaceae, Hippuridaceae, Hydrostachyaceae and Phrymaceae) the highest (100). Later, Sporne[490] used more sophisticated methods of analysis based upon a more up-to-date classification, and listed 30 main pairs of correlated characters, of which 21 were in the earlier list (Table 2.1). By this method Magnoliaceae (25) were still almost lowest (three families each with only one genus scored 23 or 24) but Dipsacaceae (87) became the highest.

There is broad agreement between Sporne's set of characters and those of Hutchinson and Bessey (Table 2.1), although in fact only eight of Sporne's 26 characters were used by Bessey or Hutchinson, and in one of these (flowers bisexual/unisexual) there is complete disagreement over the direction of evolution. The only other character that this applies to is the leaf arrangement (spiral/opposite), in which case the odd man out (Bessey) is surely wrong. On the basis of all three sets of characters in Table 2.1, the Magnoliaceae and relatives come near the 'primitive' end of the sequence, and families such as the Valerianaceae and Asteraceae near the 'advanced' end. Striking confirmation of the ancientness of Magnolioid characters in the fossil record has recently been provided by the work of various palaeobotanists,[61,300] although other groups such as the Amentiferae in fact appeared in the pollen record at about the same time. Nevertheless, there is now virtually universal acceptance of the Magnolioid families as the most primitive angiosperms (although not necessarily as representing their prototype or archetype), and of the Amentiferae as a more derived (reduced), polyphyletic group of rather diverse relationships.[437]

It must be remembered, however, that the catalogue of increasing advancement index is in no way a linear sequence of evolution, and is not a direct means of obtaining a phylogenetic classification. It is best used, if at all, in improving existing classifications and in trying to settle arguments over particular points.

Many phylogeneticists have illustrated their opinions not only in new classifications, but also by the construction of charts or 'family trees' which show the purported lineage of existing families. These are mostly drawn as dendrograms, like that representing the hierarchy in Fig. 1.1A, but there is great variation in the manner in which they are depicted; Figs 2.7 and 2.8 show two examples. Dendrograms that are supposed to represent phylogenies are variously called *phylogenetic trees*, *cladograms* or *phylograms*, but, in view of the recent adoption of the term 'cladogram' to describe the results of cladistic analyses (q.v.), that term is better dropped in favour of 'phylogram' for the purposes under present discussion. Dendrograms based solely on phenetic data are known as *phenograms*. In a phylogram the branching pattern along the vertical axis represents ancestry in *time*; in a phenogram it represents the *pattern* of phenetic relationship (and in a cladogram the *pattern* of ancestral relationship).

Sporne has, on the other hand, represented his conclusions by a bird's-eye view of a phylogram drawn in a circle, where the radius represents the advancement index and the position on the circumference the likely degree of divergence (Fig. 2.9). Dahlgren[89] originally also presented his data in the form of a phylogram in which the living orders were all placed at the topmost plane

Table 2.1 Comparison of characters alleged to be primitive/advanced by Bessey,[34] Hutchinson[217] and Sporne.[400,490] Characters over which there is positive disagreement are italicized.

Bessey	Hutchinson	Sporne
Chlorophyll/no chlorophyll	Chlorophyll/no chlorophyll	
	Terrestrial/aquatic	
Woody/herbaceous	Woody/herbaceous	Woody/herbaceous
	Trees/climbers	
	Perennials/annuals	
Stem simple/branched		
Bundles collateral/scattered	Bundles collateral/scattered	
Leaves simple/compound	Leaves simple/compound	
Leaves evergreen/deciduous		
Leaves opposite/spiral	*Leaves spiral/opposite*	*Leaves spiral/opposite*
		Leaf-margins toothed/entire**
		Stipules present/absent
		Secretory cells present/absent
Venation reticulate/parallel		
		Wood unstoreyed/storeyed*
		End-plates scalariform/not
		Side-walls scalariform/not
		Rays heterocellular**
		Parenchyma apotracheal/paratracheal
		Leuco-anthocyanins present/absent
		Ellagitannins present/absent
		Aluminium accumulated/not*

Flowers solitary/clustered		
Floral parts spiral/whorled		
Flowers polymerous/oligomerous	Flowers polymerous/oligomerous	
Flowers bisexual/unisexual	*Flowers bisexual/unisexual*	*Flowers unisexual/bisexual*
Flowers monoecious/dioecious	Flowers monoecious/dioecious	
Flowers petalous/apetalous	Flowers petalous/apetalous	
Flowers actinomorphic/zygomorphic	Flowers actinomorphic/zygomorphic	Flowers actinomorphic/zygomorphic
Petals free/fused	Petals free/fused	Petals free/fused**
		Petals or sepals numerous/few**
		Petals or sepals imbricate/overlapping**
Stamens many/few	Stamens many/few	Stamens many/few
Stamens separate/fused	Stamens separate/fused	Stamens separate/fused*
Pollen powdery/massed		
		Tapetum glandular/amoeboid
		Pollen 1- or 2-/3-nucleate
		Pollen few/many pored
Flowers hypogynous/epigynous	Flowers hypogynous/epigynous	Flowers hypogynous/epigynous**
Gynoecium polycarpous/oligocarpous	Gynoecium polycarpous/oligocarpous	Gynoecium polycarpous/oligocarpous
Gynoecium apocarpous/syncarpous	Gynoecium apocarpous/syncarpous	Gynoecium apocarpous/syncarpous*
		Placentation axile/parietal*
		Integuments 2/1 or 0
		Integuments with/without bundles
		Ovules crassinucellate/tenuinucellate
		Ovules anatropous/not**
		Seeds >1/1 per carpel**
	Fruits single/aggregate	
	Fruits capsule/drupe or berry	
		Seeds arillate/not
Seeds with/without endosperm	Seeds with/without endosperm	Seeds with/without endosperm**
		Endosperm nuclear/cellular

* excluded from 1980 list
** added to 1980 list

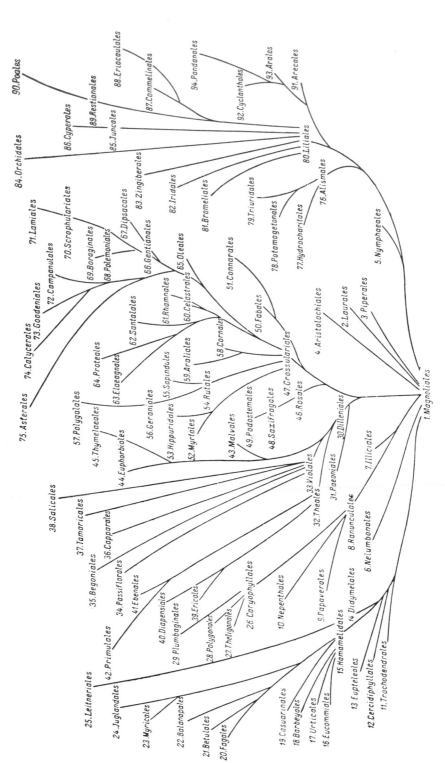

Fig. 2.7 Phylogram of the orders of angiosperms taken from the 1966 version of Takhtajan's scheme of classification.

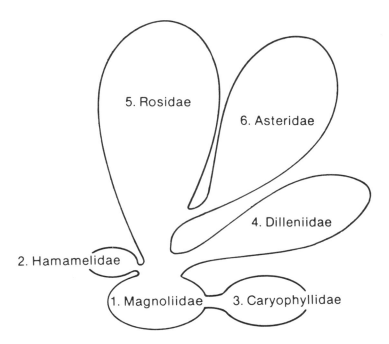

Fig. 2.8 Phylogram of the six subclasses of dicotyledons, taken from Cronquist[63].

of the diagram; the ancestral branches (absent in Sporne's diagram) all represent unknown taxa, many of which are extinct (Fig. 2.10). Proximity of groups on the topmost plane is defined by their degree of phenetic resemblance. However, in all these pictorial representations the exact shapes of taxa, the points of branching, and the lateral or circumferential order and spacing have no precise meaning—they are merely aids to the eye. In realization of this, Dahlgren's later schemes have dropped the branching portion of the phylogram and use only the topmost plane;[91,92] similar representations are used by Thorne.[438]

In the most modern schemes, as typified by that of Dahlgren, it is accepted as unlikely that many modern living groups of angiosperms are the ancestors of other living groups. Thus the Magnoliales are not seen as an ancestral group, but as a group that has retained a high proportion of ancestral characters. Secondly, the characters used in classification have been drawn consistently from a wide field of biology; in particular anatomical, developmental and chemical characters figure very prominently. In Dahlgren's latest scheme[92,93] the dicotyledons are divided into 25 superorders and the monocotyledons into 10; Thorne's latest scheme[438] has 19 and 9 superorders respectively. Cronquist's latest scheme[77] groups orders more conservatively into 6 and 5 subclasses of dicotyledons and monocotyledons respectively, while Takhtajan's latest scheme[436] uses both subclasses and superorders—7

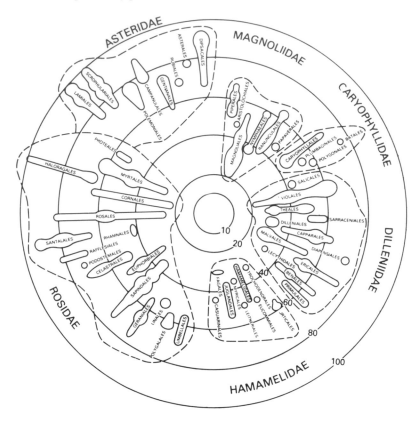

Fig. 2.9 Circular representation of the phylogenetic classification of the dicotyledons redrawn from Sporne.[399]

and 20 respectively in the dicotyledons and 3 and 8 respectively in the monocotyledons. In the last three schemes the Magnoliaceae and Asteraceae are placed respectively in the first and last major groupings within the dicotyledons, but there is less agreement in the case of the monocotyledons. Dahlgren's scheme[92] is still more different, but again the Magnoliaceae are in the first major grouping.

In emphasizing the distinction between phenetic and phylogenetic relationships, Huxley[221] coined two terms: the *grade*, a level or stage of evolutionary advancement; and the *clade*, a single phylogenetic lineage. Obviously a clade may contain many grades, and a grade many clades—grades and clades are simply the horizontal and vertical entities respectively seen in a phylogram—but in a phylogenetically based classification each taxon represents a single clade (i.e. it is monophyletic), and the lowest taxonomic rank is the finest grade of a clade.

The most direct evidence for phylogenetic pathways, i.e. the existence of fossils linking the major modern taxa, or showing a different spectrum of taxa in prehistoric times from now, is mainly missing for the angiosperms, and the

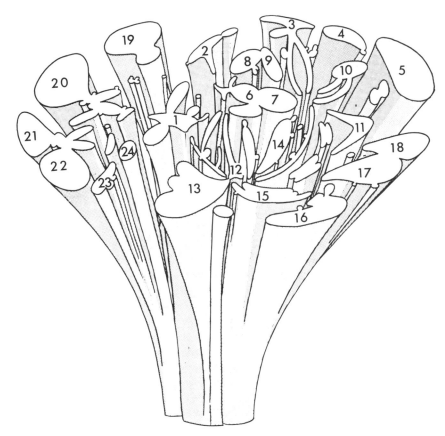

Fig. 2.10 Phylogram of the orders of angiosperms representated as the transection of an imaginary phylogenetic tree, adapted from Dahlgren.[89] Only 24 of the largest orders are named, as indicated by the following numbers: 1, Magnoliales; 2, Ranunculales; 3, Rutales; 4, Araliales; 5, Asterales; 6, Malvales; 7, Euphorbiales; 8, Violales; 9, Capparales; 10, Santalales; 11, Solanales; 12, Rosales; 13, Fabales; 14, Myrtales; 15, Ericales; 16, Gentianales; 17, Scrophulariales; 18, Lamiales; 19, Caryophyllales; 20, Orchidales; 21, Cyperales; 22, Poales; 23, Arecales; 24, Arales.

classifications discussed above have therefore been constructed by indirect means. In the case of pteridophytes and gymnosperms, on the other hand, there are many relevant data, and ideas on the classification of these groups have changed radically in relatively recent times. The history of fern classification has been summarized by Pichi Sermolli.[325] In both groups the major advance in phylogenetic studies has been the realization, largely through the study of fossils, that the Pteridophyta and Gymnospermae are not so much *taxa* as similar levels of organization and that both are polyphyletic assemblages of taxa (grades rather than clades). Thus most modern schemes of classification do not recognize either of the above two groups as taxa, but instead recognize about four or five taxa of pteridophytes (e.g. Psilophytina, Lycophytina, Sphenophytina, Filicophytina) and three to seven taxa of gymnosperms (e.g. Cycadopsida, Coniferopsida, Gnetopsida), each roughly on a level (subdivision or class) with the whole of the angiosperms. Many examples repeating this sort of process at lower levels within the two groups could be given.

Phylogenetically useful fossils among the non-vascular plants are largely unknown, and we have no real idea of their pathways of evolution. Modern classification of algae and bryophytes is, therefore, strictly phenetic rather than phylogenetic, although some taxonomists would consider it phylogenetic as well. The Bryophyta have retained their entity in all major systems of classification, but the algae, even before the pteridophytes and gymnosperms, became split into a number of different divisions (see Appendix), each considered to be monophyletic. Some biologists have, for the same reasons, preferred to abandon the Thallophyta as a formal taxonomic group. Whether or not names such as Thallophyta, Pteridophyta and Gymnospermae are considered taxonomic groups, they are useful descriptive nouns which can be used non-committally as thallophytes, pteridophytes, gymnosperms, etc.

Many of the supposedly phylogenetic overall plant classifications which have been proposed in the past half century are usefully surveyed by Edwards.[109] Almost inevitably his discussion leads to yet another new classification, thought to be better (i.e. more phylogenetic) than any of its predecessors; as intimated above, whether or not this is so is unknown. Further aspects of the classification of lower plants are discussed in Chapter 8.

At the lower levels of the taxonomic hierarchy more direct evidence of evolution is available through the methods of cytology and genetics, which rose to the forefront of taxonomy (as biosystematics or 'the new systematics') in the 1940s, 50s and 60s (see Chapter 6). At and around the species level it has proved possible in many genera to establish without any reasonable doubt the precise phylogenetic pathways. Before the age of phylogenetic classifications the variety was about the only infraspecific rank used. During the preparation of phylogenetic classification the extensive use of other ranks, notably the subspecies, and various redefinitions of the species, have sprung from the results of cytogenetic and genecological studies; and logically so, for the unit of evolution, the breeding population, is the same as the object of those studies.

From the vast amount of effort that has been put into the production of phylogenetic classifications in the last hundred years, during which time micromorphology, genetics, cytology, biochemistry and other topics generally gathered under the title of cell biology have emerged and been utilized by taxonomists, and our knowledge of fossil plants has vastly improved, it might be imagined that great changes in our systems of classification have taken place. In general this is not so, either in the higher levels of the hierarchy or in the lower levels. This is partly because we have not been able to unravel with any degree of certainty or detail the major phylogenetic pathways of plants, as mentioned previously, and partly because our systems of classification based on purely phenetic criteria broadly agree with those which we have been able to base on phylogeny. There is, in fact, a general belief that a perfect phenetic system would also be a perfect phylogenetic one. It seems very unlikely that they will be extremely different (presumably all members of the Poaceae, for example, *are* phylogenetically related), but we do not know to what extent the evolutionary processes of convergence, parallelism and catastrophic divergence have distorted the phylogenetic pattern as shown by the present-day phenotype. The general similarity of phenetic and allegedly phylogenetic systems of classification may also be due to the fact that we have attempted to express our phylogenetic data by use of exactly the same hierarchy of taxonomic ranks and the same names as we use to express our phenetic data. Perhaps this imposes too severe strictures to allow for radical change. Whatever the cause, the differences between Bentham and Hooker's and Cronquist's classifications of angiosperms are those of detail rather than of outline. This concordance is perhaps not so surprising if one assumes that evolution has in general progressed along lines of greater specialization and complexity, for phenetic classifications have mostly started with the simple and progressed to the complex. Instances where big changes in the classification of higher plants have been made in the light of phylogenetic data very often involve taxa whose simplicity of structure has come to be regarded as reduced (specialized or advanced) instead of primitive, for example loss of petals, reduction in stamen or carpel number.

In the lower plants there have certainly been great taxonomic changes, which might at first be seen as the result of advance in phylogenetic understanding. In fact, for the most part, these changes have had a quite different origin. An intimate knowledge of most lower plants is possible only by their study with modern biochemical and microscopic aids; their variation is at a lower order of magnitude of size than that of higher plants, and in the nineteenth century the means to study the variation adequately did not exist. Had the pre-Englerian taxonomists had access to the techniques of electron microscopy, chromatography, spectrometry and the like, there is little doubt that phenetic classifications of lower plants broadly similar to the present-day ones would have been produced.

Phase 7. Modern phenetic methods (taxometrics)

The many difficulties associated with producing phylogenetic classifications of plants, and the uncertainties inherent in those that were produced, led in the 1950s and 1960s to a new methodology in classification, whereby a totally phenetic approach was advocated. Some of the strongest arguments for this line have come from Heywood[204] and Raven,[343] who believe that phenetic classifications represent a more practical solution than phylogenetic ones. The beginning of this school of thought coincided with the independent introduction of the methods and concepts of *numerical taxonomy* or *taxometrics* (taximetrics, phenetics) by Sneath[390,391] and Michener and Sokal,[293] applied respectively to bacteria and bees. As a result of this coincidence numerical taxonomy has come, in many eyes, to be almost synonymous with phenetic classification. It must be realized, however, that this is not the case. Numerical taxonomy does not produce new data and is not a new system of classification, nor even a new set of principles underlying one, but rather a new method of organizing data and obtaining from them a classification (if that be desired) or some other form of presentation. There are other ways than taxometric ones of producing a modern phenetic classification, and other aims and uses of modern numerical taxonomy than the production of a phenetic classification. Nevertheless, in this book the term taxometrics is used only in relation to phenetic taxonomy.

The independent researches of **P. H. A. Sneath** and **R. R. Sokal** led to their collaboration and production of a standard textbook, *Principles of Numerical Taxonomy*,[394] which after ten years was entirely rewritten and updated as *Numerical Taxonomy*.[393] The existence of these volumes renders unnecessary the formulation in this book of much detail concerning their subject, for *Numerical Taxonomy* is a thorough treatise explaining not only the methods and results but also the theoretical background, aims and principles. Seven main advantages (of which the authors are utterly convinced) of numerical taxonomy over conventional taxonomy are given, and seven main principles are laid down (comments are added in brackets):

1. The greater the content of information in the taxa of a classification and the more characters on which it is based, the better a given classification will be (better in this context can be equated with 'more predictive').
2. *A priori*, every character is of equal weight in creating natural taxa.
3. Overall similarity between any two entities is a function of their individual similarities in each of the many characters in which they are being compared.
4. Distinct taxa can be recognized because correlations of characters differ in the groups of organisms under study.
5. Phylogenetic inferences can be made from the taxonomic structure of a group and from character correlations, given certain assumptions about evolutionary pathways and mechanisms (this principle was not one of the original (1963) ones).
6. Taxonomy is viewed and practised as an empirical science (as opposed to an interpretative or deductive science).
7. Classifications are based on phenetic similarity (but note the possibility under 5 above).

Most of these principles (except 5) bear resemblance to the aims and methods of Adanson, (see Phase 5 of this chapter), and are therefore often known as ***neo-Adansonian principles***. Inasmuch as it is useful to remind oneself that these ideas are not new, and that they were once held in a very different academic environment from that of today, this term is useful, but it is hardly accurate to visualize Adanson as the founder of numerical taxonomy; moreover, the classifications produced by him were inferior (on any basis) to those produced by several of his contemporaries using non-empirical methods.

These seven principles underlie not only Sneath and Sokal's concepts of numerical taxonomy, but the subject as a whole, and nowadays essentially the same aims and methods are being applied to all groups of microorganisms, animals and plants.

The basis of numerical taxonomy is the logical corollary to the fact that special purpose classifications are based on one or few characters or on one set of data; it seeks to base classifications on a great number of characters from many sets of data (organs, methods of investigation, groups of subst-ances, etc.) in its bid to produce an entirely phenetic classification of maximum predictivity. All the characters so used are *a priori* weighted equally (i.e. no weighting) according to the methods advocated by Sneath and Sokal, despite claims by some that weighting should be attempted and the existence now of methods of achieving it *a posteriori*. Sneath and Sokal argue that, even if some characters *should* receive a greater or lesser weighting, such considerations will be more or less nullified because of the large number of characters used. For the same reason, the deliberate, forced or unwitting omission of characters from the exercise (***residual weighting***), even ones known to be useful taxonomically in the group under study, is theoretically not a problem. These methods should also solve arguments, discussed above, of whether two similar taxa are truly closely related or exhibit convergence or parallelism, because use of a very wide range of characters will prevent too much weight being placed on a few conspicuous ones.

The number of characters needed to achieve this desirable situation has been much discussed, and no absolute answer can be given. It is widely regarded that 60 characters should be considered a minimum, and that 80 or 100 (or more) are desirable, and this provides a reasonable working basis. There is also the problem of what constitutes a character; indeed, one of the most valuable aspects of numerical taxonomy has been the need that it has imposed on taxonomists to reconsider the precise meaning of many of their terms (as well as their aims). One might consider petal-shape a character, yet on closer inspection the shape is clearly the summation of several different variables, for example length, breadth, apical notching, basal attenuation; all of these are separate characters, and it is usually quite easy to amass a list of 80 or more characters for a given plant. As Sibley[379] has said: 'one person may treat a structure as one character while another applies calipers to a dozen different dimensions of the same structure and emerges with 12 characters'.

In listing characters it is of course important that only ***homologous*** characters be compared. Many obvious pitfalls are well known, for example the 'petals' of *Anemone* and *Ranunculus* are not homologous (the former are considered to be homologous with the sepals of *Ranunculus*, true petals being absent from *Anemone*), nor are the 'fruits' of *Fragaria* (a fleshy receptacle with many small, pip-like fruits), *Rubus* (a mass of fleshy fruits) and *Prunus* (a single fleshy fruit). In most cases, however, one does not know whether one is comparing homologues, for example red flower colour, leaf pubescence, woody habit. Homology is usually defined on the basis of common evolutionary origin, a definition which should in theory be uncontentious, but which in fact is usually quite impractical because of our lack of evolutionary data. In practice, therefore, one can often only guess at homologies by making as detailed as possible an investigation of the structures concerned.

The basic unit of numerical taxonomy is the ***operational taxonomic unit (OTU)***, which is the term given to the lowest taxon being studied in a particular investigation. Hence the OTUs that one is classifying might be families, species or individuals, or any other taxonomic entity. Once one has chosen the OTUs to classify, and the characters by which they are to be classified, each OTU has to be scored for the possession of one or other ***character-state*** or ***attribute*** for each character. This results in a ***data matrix*** of attributes (OTUs × characters, or $t \times n$), as shown in Fig. 2.11. If one is classifying 30 OTUs and using 100 characters, the data matrix will consist of

OTUs (t) \ Characters (n)	a. Epidermal long-cells wavy	b. Inflorescence spiciform	c. Glumes subulate	d. Lower glume minute	e. Lower glume twisted	f. Lemma long-awned	g. Lemma bifid	h. Lemma keeled	i. Lemma tuberculate	j. Lemma 3-veined	k. Lemma broadly hyaline	l. Hilum punctiform
1. Castellia	−	±	−	−	−	−	−	−	+	−	−	−
2. Catapodium	−	±	−	−	−	−	−	±	−	−	±	+
3. Ctenopsis	+	+	−	+	−	−	−	−	+	−	−	−
4. Cutandia	−	−	−	−	−	−	±	+	−	+	±	−
5. Desmazeria	−	±	−	−	−	−	−	+	−	±	±	+
6. Loliolum	−	+	+	−	+	−	−	−	−	−	−	−
7. Micropyrum	+	+	−	−	−	±	±	−	−	−	−	−
8. Narduretia	+	−	−	+	−	+	−	−	−	−	−	−
9. Narduroides	+	+	−	−	−	−	+	−	−	−	+	+
10. Vulpia	+	−	−	±	−	+	−	−	−	−	−	−
11. Vulpiella	−	−	−	−	−	±	+	+	−	+	−	−
12. Wangenheimia	+	+	+	−	+	−	−	+	−	−	−	−

Fig. 2.11 Small data matrix involving 12 OTUs (t) and 12 characters (n), and hence 144 attributes (t × n). The OTUs are genera of annual grasses in the tribe Poeae. Possession of a character-state is marked by +, lack by −, presence or absence in various species by ±.

3 000 attributes; clearly, it is not difficult to become involved with large numbers of pieces of information. For this reason, and because of the number and variety of operations needed to be carried out on these data, a computer is virtually essential.

The use of a computer necessitates the codification of the attributes in some simple form which can be fed into the computer, and this often presents great problems. The simplest codification is a **binary** or **two-state** system, for example + and − (as used in Fig. 2.11), or 0 and 1, where each character can exist in only two states. One might code for leaves hairy versus leaves glabrous, or petals coloured versus petals white, etc. But of course a great many characters are not of this sort. They may be **multistate qualitative** characters (e.g. petals white, yellow, red, blue) or **multistate quantitative** characters (e.g. pubescence measured by the number of hairs per unit area, or redness by the amount of pigment per unit volume). Since the great majority of taxometric procedures demand, or are greatly facilitated by, the use of only binary characters, it is customary to convert multistate characters into the latter. This is not difficult, but in many cases it involves a good deal of subjectivity. For example, leaf-length can be simply converted to leaves long versus leaves short, but one has to decide the numerical range of these two attributes. Petals white, yellow, red or blue can be converted to petals white versus petals coloured (in which case the differences between red and yellow, etc., are lost), or to a series of characters such as red versus not red, yellow versus not yellow, etc. (in which case some might argue that colour is not being given equal weighting to, say, leaf colour, which might be only ever green or purple). This form of subjectivity subverts the objectivity of numerical taxonomy, although, perhaps, the large number of characters used will again neutralize any such defects.

Other sorts of problems also occur. In a group of 30 OTUs, one might score for petals white versus petals red. What, however, if one OTU lacks petals, or if one OTU is known only in a fruiting stage, so that (in either case) the character is inapplicable? Obviously the simplest way out is to omit that character, or perhaps that OTU, from the matrix, but this might not be desirable. In other cases a character-state might vary within an OTU. If an OTU is a species, some individuals might have red and others white petals; if an individual, some branches might have longer leaves than others. Mean figures could be used, but this would obscure differences between the ranges of variation exhibited by each OTU (e.g. red or white versus pink).

Despite difficulties such as the above, it is usually possible to construct an acceptable data matrix of two-state attributes for the computer program, of which many variants are now available. Basically, the computer sorts out (**clusters**) the OTUs according to their overall similarity, i.e. according to the number of attributes in common. Such a process is termed **cluster analysis** and is the usual method of analysis employed in taxometrics. Usually this is achieved not by direct use of the data matrix, but via the production of a table of similarity or dissimilarity coefficients, which is a $t \times t$ table (**similarity** or **dissimilarity matrix**) giving a measure of similarity or difference to all possible combinations of pairs of OTUs (Fig. 2.12).

The results of cluster analysis are often visualized as a hierarchical dendrogram of phenetic relationships (phenogram) (Fig. 2.13A), in which less and less similar OTUs are successively linked together. Alternatively various non-hierarchical versions may be prepared (Fig. 2.13B), the OTUs being linked by straight lines of different lengths or by contour-like lines encircling different numbers of OTUs. These two methods are analogous to the elevation and plan views of the hierarchy in Fig. 1.1. There are many other systems of representation besides these, for example shaded similarity matrices and various sorts of three-dimensional diagrams or models. A useful survey of types of cluster analysis is given by Abbott *et al.*[1]

	1	2	3	4	5	6	7	8	9	10	11	12	13	14	15	16	17	18	19	20	21
ameghinoi 1	x	14.4	16.9	22.5	19.9	13.1	24.5	21.1	14.9	11.2	30.7	9.1	9.0	30.3	4.3	26.4	16.3	5.7	24.4	21.0	22.4
baccharoides 2		x	31.8	37.4	5.0	28.0	39.4	36.0	29.3	26.1	45.6	5.8	15.3	15.4	10.6	41.3	31.2	20.6	9.5	35.8	31.2
bracteata 3			x	5.6	36.0	7.6	17.0	4.2	9.4	5.7	13.8	26.0	25.9	46.4	21.2	9.5	10.8	11.2	40.5	13.5	14.9
californica 4				x	42.4	13.2	24.6	9.8	15.0	11.3	19.4	31.6	31.5	52.8	26.8	15.1	16.4	16.8	46.9	21.1	22.5
espinosae 5					x	33.0	44.4	41.0	34.8	31.1	50.6	10.8	20.3	10.4	15.6	46.3	36.2	25.6	4.5	40.9	42.3
gillesii 6						x	15.2	11.8	5.6	1.9	21.4	22.2	22.1	43.4	17.4	17.1	7.0	7.4	37.5	11.7	13.1
gluttinosa 7							x	23.2	9.6	13.3	32.8	33.6	33.5	54.8	28.8	28.5	18.4	18.8	48.9	3.5	17.1
grandis 8								x	13.6	9.9	9.6	30.2	30.1	51.4	25.4	5.3	15.0	15.4	45.5	19.7	21.1
iserni 9									x	3.7	23.2	24.0	23.9	45.2	19.2	18.9	8.8	9.2	39.3	6.1	7.5
mandonii 10										x	19.5	23.0	20.2	41.5	15.5	15.2	5.1	5.5	35.6	9.8	11.2
microcephala 11											x	39.8	39.7	61.0	35.0	4.3	24.6	25.0	55.1	29.3	30.7
neaeana 12												x	9.5	21.2	4.8	35.5	25.4	14.8	15.3	30.1	31.5
repens 13													x	30.7	4.6	35.4	25.3	14.7	24.8	30.0	31.4
resinosa 14														x	26.0	56.7	46.6	36.0	5.9	51.3	52.7
ruiz-lealii 15															x	30.7	20.6	10.0	20.1	25.3	26.7
sarothrae 16																x	20.3	20.7	50.8	25.0	26.4
serotina 17																	x	10.6	40.7	14.9	16.3
spathulata 18																		x	30.1	15.3	16.7
talalensis 19																			x	45.4	46.8
texana 20																				x	13.6
gayana 21																					x

Fig. 2.12 Dissimilarity matrix, showing a measure of the phenetic differences between all possible pairs of all 21 species and subspecies of the genus Gutierrezia (Asteraceae), reproduced from Solbrig.[397] The higher figures indicate greater *dissimilarity*.

These different methods simply seek to present the results of the analysis in a form which can be comprehended by the human eye and brain and converted into a classification. They represent the *taxonomic structure* of the group of OTUs under study, that is the disposition of the OTUs in an imaginary taxonomic, multi-dimensional space or *hyperspace*.

The clusters of OTUs therein recognized are defined by the possession of the greatest number of shared features, i.e. they are *polythetic*. None of these features is individually either necessary or sufficient to define the group, i.e. lack of any feature by a taxon does not exclude it from the group. *Monothetic* groups, on the other hand, are defined by the possession of a unique set of features which is both necessary and sufficient to define the group. Monothetic groups can easily be constructed, but they are usually (though not invariably) artificial, e.g. Linnaeus' class Monandria.

The limits of the taxa may be decided (to use one example) by drawing horizontal lines at various positions on the phenogram (Fig. 2.13A). The decision of which lines to take as those delimiting various taxa of different levels is, however, a major problem in numerical taxonomy, as it can be in classical taxonomy. Traditionally, taxa are derived not by deciding *a priori* on different points of delimitation (i.e. basing them on their degree of internal diversity), but by taking their limits as coinciding with the points of greatest disjunction from other taxa. This results in taxa not only of very different sizes, but also of very different degrees of diversity. This is equivalent to choosing *a posteriori* different levels of the phenogram in different groups. Although some numerical taxonomists might not approve of the limits of taxa being drawn subjectively by choosing the most convenient discontinuities, there is no way yet of equating different levels in the phenogram with the different taxonomic ranks (family, genus, etc.).

This problem also gives rise to the question of whether there is any equivalence between the various ranks in different groups of organisms. Is a family of flowering plants, for example, equivalent in any sense to one of algae or of insects, or of fish? Theoretically, numerical taxonomy gives us the opportunity to answer this question, but so far it has not succeeded, and hopes that work on protein or nucleic acid sequences will one day do so are far from being realized. In fact the question is probably not meaningful, for it is difficult to see how one could judge whether two fish are more or less similar to each other than are two trees. The question can only be answered in relation to a narrow spectrum of characteristics; for example, the fact that hybrids have been obtained between fish from different orders indicates that, in terms of breeding barriers, the orders of fish are far more closely related than those of flowering plants (where not even inter-familial hybrids are known), but one cannot extrapolate this fact to other sets of characters (e.g. protein similarities).

For this reason a new terminology has been advocated by some numerical taxonomists 'for the hierarchical system established by numerical taxonomy'.[393] In this, the term *phenon* replaces *taxon*, i.e. a rank of any level, and the particular phenons (note plural) are designated by numerical prefixes showing the level of resemblance by which they are defined. One can thus talk of 50-phenons or 70-phenons, etc.; these refer to phenons defined by a 50% (or 70%) similarity in the attributes over which they were scored, and they

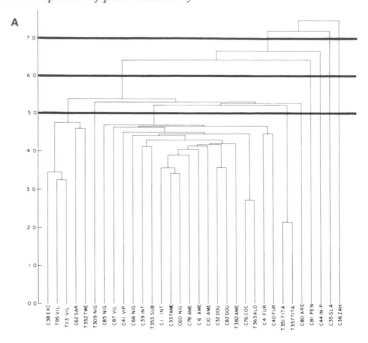

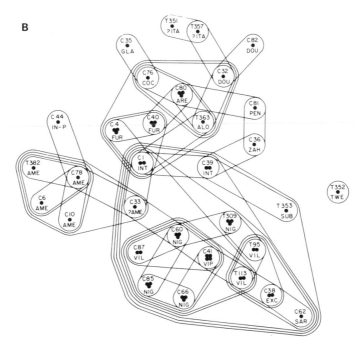

Fig. 2.13 Two ways of expressing the results of cluster analysis: **A,** a phenogram; **B,** non-hierarchical cluster diagram. Both are taken from Edmonds[108] and refer to 32 populations of *Solanum*. In **A** the vertical axis is the measure of dissimilarity, rising to a theoretical maximum of 29; phenon lines have been drawn in at levels 5, 6 and 7.

are physically delimited by drawing horizontal straight lines (***phenon lines***) at the appropriate level across a phenogram. In Fig. 2.13A phenon lines have been drawn across arbitrarily at the 5.0, 6.0 and 7.0 levels, these delimiting respectively 9, 5 and 3 taxa. It should be emphasized that, whereas the *pattern* of phenons should be the same whatever analysis is undertaken, the absolute numerical value of particular phenons will vary according to the method of analysis; 'They are therefore comparable only within the limits of one analysis'.[393]

A point which should be considered is whether the pattern of phenons is in reality the same whatever analysis is undertaken, as suggested above. There are two main variables to be considered here, the type of computer program adopted and the nature of the characters scored. In both cases it must be admitted that, although a reasonably similar classification usually results, in some cases it does not. This is often so when one compares classifications made upon different parts of plants, only one of which might be available, for example fragmentary fossils or rarely collected plants, or different stages of the life cycle. The fossil group known as the seed-ferns or pteridosperms, for example, might be classified along with pteridophytes on the basis of vegetative structure, but with gymnosperms on the basis of reproductive organs. There are many other cases among fossil plants where two elements were classified quite differently, until they were found to be physically connected. *Ginkgo*, vegetatively close to the conifers but reproductively closer to the cycads, is a living example, and it is well known that the classification of various algae and bryophytes differs according to whether emphasis is placed on the sporophyte or on the gametophyte. In certain ferns the gametophyte asexually reproduces independently from the normal sexual life-cycle, and the gametophytes have in some cases come to occupy a different area of distribution from the sporophytes.[149] Presumably they are genetically different as well. This problem, termed by Sneath and Sokal *incongruence* of classifications, is not peculiar to numerical taxonomy but applies to taxonomy in general. Nevertheless, there is evidence that the large number of characters used in numerical taxonomy does not wholly rule out the need to ensure a wide and even distribution of characters over the organisms being studied, and the partial incongruence of classifications using the same data but different numerical techniques adds to the problem. An advantage of numerical taxonomy here, however, is that incongruence can be accurately measured.

In the preceding paragraphs many difficulties and problems concerning practical numerical taxonomy have been briefly discussed, and perhaps these aspects have been emphasized somewhat at the expense of the positive ones. All these problems can, theoretically, be resolved in time, and perhaps they will be; Sneath and Sokal say that they amount to 'largely technical challenges that will be overcome before very long'. For instance, computer programs are already available that will introduce an element of *a posteriori* weighting, or will combine codification of two-state and multistate characters. Some taxonomists even 'anticipate in the near future that automated methods will be introduced for extracting information directly from specimens and converting it into unit characters'.[393]

In the view of the writer, it is most unlikely that numerical taxonomy will ever supplant traditional methods as the standard procedure. It does not seem destined to *replace* other techniques, nor to produce a type of classification that *replaces* others. Numerical taxonomy is likely to be most successful where other methods have failed, or are laborious, or otherwise difficult to apply, for example with various microorganisms[392] and the elms (*Ulmus*).[360] A good example of its use at the family level and above is the re-classification of the dicotyledons by Young and Watson,[485] who used 83 attributes (all morphological or anatomical) in their analysis of 543 representative genera. One notable conclusion was that the major division of OTUs was between crassinucellate genera (ovules with a thick nucellus) and tenuinucellate ones (with a thin nucellus). (This is one of Sporne's 26 major characters).

The view put forward here is that taxometrics is in the foreseeable future likely to be most useful in examining existing classifications, assessing their good and bad features, and in some cases pointing to ways of modifying and improving them. So far, its main value has been in teaching taxonomists to analyse their aims, methods, data-collection and -presentation, and conclusions more logically and objectively. An excellent illustration of this is the work on grass classification at the levels between family and genus carried out by L. Watson and colleagues,[466,467,468] who have amassed an enormous amount of data (over 300 characters) on over 720 genera (virtually all of the world's total) and are using this information to improve the classification of the family.

Phase 8. Modern phylogenetic methods (cladistics)

W. Hennig's textbook[193] entitled *Phylogenetic Systematics*, published in 1966 but based upon his earlier textbook *Grundzüge einer Theorie der phylogenetischen Systematik* (1950), effectively founded the subject known as phylogenetic systematics, now more usually termed **cladistics** (a term coined in 1969 by E. Mayr). Its earliest proponents (cladists) were zoologists, but since the mid-1970s many botanists (notably in the USA) have adopted its principles. A separate but convergent approach to cladistics came in the form of **W. H. Wagner**'s *groundplan-divergence method*,[456] which can be traced back to Wagner's work on Hawaiian ferns in 1952. Certainly no subject can rival cladistics for the degree of controversy and argument (too often acrimonious and personal) that it has brought to taxonomy. The literature of cladistics is already voluminous and is growing rapidly, and only a brief over-simplified survey comparable with that of taxometrics can be given here. Many of the most important developments are published in the journals *Systematic Zoology*, *Systematic Botany* and the newly formed *Cladistics* (Volume 1 in 1985). In addition there are several conference reports, notably those of the Willi Hennig Society entitled *Advances in Cladistics* (Volume 1 in 1981), textbooks,[1,54,115,304,477] and reviews.[49,50,103,105,119,144,146,214]

Cladistics is basically a methodology that attempts to analyse phylogenetic data objectively (and hence to produce an objective phylogenetic classification), in a manner parallel to that in which taxometrics seeks to introduce objectivity into phenetics and phenetic classifications. Despite this similarity, the two approaches and their respective proponents are largely in conflict.

Moreover, there is also much argument between cladists concerning differing aims, methods and taxonomic interpretations.

The methods of Wagner, Hennig and many later cladists are often gathered under the heading of *parsimony methods*, in that they can be said to utilize the *principle of parsimony*, whereby the shortest hypothetical pathway of changes that explains the present phenetic pattern is considered to be the most likely evolutionary route. These cladistic methods are not the only ones to adopt the principle of parsimony, for the latter has been used by phenetic taxonomists,[44,59] who have attempted to relate extant OTUs without hypothesizing *a priori* any ancestor or directions of evolutionary change. The above cladistic methods differ fundamentally from taxometrics in that deductive (*a priori*) reasoning is used to determine routes of evolutionary change.

Hennig differentiated between monophyletic, paraphyletic and polyphyletic groups (see p. 30), and stipulated that taxa in a truly phylogenetic system should be only monophyletic. In order to identify monophyletic taxa, cladograms such as those in Fig 2.6 are constructed by considering characters in which a primitive (*plesiomorphous*) and an advanced (*apomorphous*) character-state can be recognized, e.g. superior versus inferior ovaries respectively. The possession of plesiomorphous character-states in common by a group of taxa is termed *symplesiomorphy*, and the possession of derived character-states in common is termed *synapomorphy*. By reference to Fig. 2.6 it can be seen that symplesiomorphy does not necessarily indicate monophyly, for it is equally indicative of paraphyly (e.g. X and Y in Fig. 2.6B). Synapomorphy is much more likely to indicate monophyly (e.g. Y and Z in Fig. 2.6A), but it can indicate polyphyly due to parallelism (e.g. X and Y in Fig. 2.6C) or convergence (e.g. X and Y in Fig. 2.6D); such cases are called *false synapomorphies*.

The ways in which plesiomorphous and apomorphous states of characters are deduced[74] causes much debate[102,279,280] and is the basis of one of the most common criticisms of cladistics. Firstly, one has to ensure that one is considering only homologous structures, as with taxometrics. Often one can be sure one way or the other, but often not, due to ignorance. For example, we can be fairly certain that the leaves of the angiosperms *Magnolia* and *Laurus* are homologous, and that those of a moss (gametophytic) and an angiosperm (sporophytic) are not; but what about the leaves of *Magnolia* and a fern (both sporophytic)?

Assuming one can restrict attention to cases of true homology, the polarity (direction) of the change in character-state is determined in various ways. Some of these ways that have been used in the past are obviously fallacious and one can only wonder why they were ever considered reliable. Foremost in this category is the tenet that 'common is primitive', i.e. that the more common character-state is plesiomorphous, the rarer one apomorphous. Some have suggested that this analysis should be carried out within the group concerned (*in-group*) and others suggest searching in relatives outside it (*out-group*). In either case there are so many obvious instances where the tenet does not hold that discussion would be superfluous. Little more useful is the ontogeny of a character, whereby it is thought that an organ, in its development to its mature character-state, often passes through evolutionarily more primitive stages, e.g. the possession of gills by foetal mammals or by tadpoles. This is use of the worn-out paraphrase 'ontogeny repeats phylogeny'; often it does, but often it does not, and hence misleads. Another criterion that can be used is that known as the 'correlation of transformation series'. Basically, it suggests that if the polarity of one character-state change (transformation) can be convincingly demonstrated, this can be extrapolated to all other character-state transformations that are correlated with it. It seems a reasonable method (so long as functionally correlated characters are recognized for what they are) but is of course only secondary, in that unequivocal evidence must be produced for at least one transformation. Strong (if not unequivocal) evidence of the polarity of a transformation might come from two sources: fossils, and *out-group comparison*. Since relevant fossils are normally not available, the latter method is in practice often the only reasonably reliable one available. Given that two character-states of a character are found in a single monophyletic group, the state that is also found in a sister-group is considered likely to be plesiomorphous and that found only within the monophyletic group under study apomorphous. In this instance the out-group being used is a *sister-group* of the group under study, i.e. another monophyletic group that shares a common direct ancestor. Two sister-groups together with their common ancestor form a larger monophyletic group, and a sister-group provides the most valuable out-group comparison. Several useful accounts of the technique of out-group comparison are available.[115,214,304,477] As well as demanding that only monophyletic groups should be recognized as taxa, Hennig proposed that sister-groups should be recognized at the same taxonomic rank.

The basic units that are manipulated in cladistics are often known as *evolutionary units* (EUs), equivalent to the phenetic OTUs. Once a set of data relating to plesiomorphous versus apomorphous character-states has been accumulated for all the EUs, a data-matrix such as that in Fig. 2.14 can be constructed. Normally included with the EUs is the hypothetical ancestor that possesses all characters in the plesiomorphous state. This matrix is then used to construct a cladogram, either directly, or indirectly (via a $t \times t$ table) as in taxometrics. If the number of EUs and characters is low this operation can be completed manually, but normally computer programs, of which a wide range is now available, are used. One aspect of the variation in the criteria used to construct the programs may serve as an example. In some cases the pairs of EUs are scored according to the number of differences in their respective character-states; this table of so-called 'Manhattan Distances' is in effect a dissimilarity matrix. In other cases the scoring is according to the number of apomorphous states in common possessed by the pairs of EUs (Fig. 2.15), ignoring the possession of plesiomorphous states in common. Since only synapomorphy is likely to define monophyletic groups, this latter method is closer to the original cladistic concept. In fact the former method can become even further modified because it is clearly not essential to specify which are the plesiomorphous and which the apomorphous states of each character, and some workers now advocate that approach. Such methods are not far from taxometrics, except that the characters used are far fewer in number, being restricted to those in which primitive and advanced states can (it is thought) be recognized.

In Wagner's groundplan-divergence method[456] the *number* of apomorphies

	Characters (n)	a. Epidermal long-cells wavy	b. Inflorescence spiciform	c. Glumes subulate	d. Lower glume minute	e. Lower glume twisted	f. Lemma long-awned	g. Lemma bifid	h. Lemma keeled	i. Lemma tuberculate	j. Lemma 3-veined	k. Lemma broadly hyaline	l. Hilum punctiform
	EUs (t)												
1.	*Castellia*	1	0	0	0	0	1	0	0	1	0	0	0
2.	*Catapodium*	1	0	0	0	0	1	0	0	0	0	0	1
3.	*Ctenopsis*	0	1	0	1	0	1	0	1	0	0	0	0
4.	*Cutandia*	1	0	0	0	0	1	0	1	0	1	0	0
5.	*Desmazeria*	1	0	0	0	0	1	0	1	0	0	0	1
6.	*Loliolum*	1	1	1	0	1	1	0	0	0	0	0	0
7.	*Micropyrum*	0	1	0	0	0	0	0	0	0	0	0	0
8.	*Narduretia*	0	0	0	1	0	0	0	0	0	0	0	0
9.	*Narduroides*	0	1	0	0	0	1	1	0	0	0	1	1
10.	*Vulpia*	0	0	0	0	0	0	0	0	0	0	0	0
11.	*Vulpiella*	1	0	0	0	0	0	1	1	0	1	0	0
12.	*Wangenheimia*	0	1	1	0	1	1	0	1	0	0	0	0

Fig. 2.14 Data-matrix of the same 12 taxa and 12 characters shown in Fig. 2.11, reclassified into putative plesiomorphous (0) and apomorphous (1) states. Genera which score ± in Fig. 2.11 have been treated as plesiomorphous with respect to that character.

is totalled for each EU to give an estimate of divergence (i.e. the vertical distance on the cladogram), and later these are interconnected according to the number of *shared* apomorphies.

The dendrograms (cladograms) that are obtained from the above programs are normally based upon the minimal (most parsimonious) way in which the EUs can be connected to account for the data in the $t \times t$ table. The cladograms are presented in various ways; two examples are given in Figs 2.16 and 2.17, but there are also several others in use. The end-points (EUs) are represented by extant taxa, as in a phenogram, and the synapomorphies can

	1	2	3	4	5	6	7	8	9	10	11	12
1	–	2	1	2	2	2	0	0	1	0	1	1
2		–	1	2	3	2	0	0	2	0	1	1
3			–	2	2	2	1	1	2	0	1	3
4				–	3	2	0	0	1	0	3	2
5					–	2	0	0	2	0	2	2
6						–	1	0	2	0	1	4
7							–	0	1	0	0	1
8								–	0	0	0	0
9									–	0	1	2
10										–	0	0
11											–	1
12												–

Fig. 2.15 Number of shared apomorphous characters between all pairs of taxa shown in Fig. 2.14, based on the data given in that Fig. Numbers along the axes refer to the genera shown in Fig. 2.14.

be variously marked on the cladogram. Whereas the nodes of a phenogram merely indicate groupings of OTUs with particular combinations of characters, because of the assumed evolutionary nature of the branching pattern of a cladogram its nodes are considered to represent ancestral monophyletic taxa. Although in some cases these may still be extant, or be known as fossils, in the great majority of cases they are wholly hypothetical and are known, together with the single hypothetical ancestor, as *hypothetical taxonomic units* (*HTUs*). They may be looked upon as hypothetical ancestral species, i.e. if we moved back in time appropriately they would be recognized as species, but today they are usually seen as supraspecific categories at various ranks.

Numerous conventions for the translation of a cladogram into an hierarchical classification have been advocated by Hennig and later workers. Insistence upon all the original conventions of cladistics has been one cause of resistance to its wider adoption,[78] and in turn has led to some cladists relaxing some of them. In particular, it seems likely that many hitherto universally recognized taxa (e.g. Bryophyta, Dicotyledonidae) are paraphyletic, and many taxonomists require more than doctrinaire statements such as 'Paraphyletic groups have no reality in nature and their use . . . is a major obstacle to progress'[49] to be persuaded to abandon them. The desire to obtain fewer

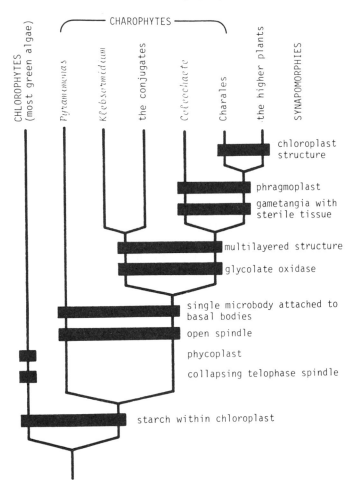

Fig. 2.16 Cladogram (rooted tree) of the major groups of green plants, taken from Bremer and Wanntorp[50], showing the character-state transformations (synapomorphies).

totally novel classifications has led to the suggested substitution of *convexity* for monophyly as a criterion of taxon recognition.[103,119,281] All monophyletic groups are convex, but paraphyletic groups (often defined by symplesiomorphies) are also convex. Convex groups may be defined as those groups in a cladogram that are wholly connected by lines that are not needed to connect taxa in a different group in the same cladogram. In Fig. 2.6B, groups YQZ and XPQY are both convex, but, in Fig. 2.6C, XY is not. In addition, not all cladists maintain that sister-groups must always be recognized at the same rank, and the extent to which the nodes of the cladogram are given taxonomic ranks also varies. Some maintain that all should be given a rank, since all are considered natural taxa. However, this introduces a practical problem in that there are often too few ranks to accommodate all the levels of nodes. This is a major problem in cladistics that has been addressed in several ways, none wholly satisfactorily. Nevertheless, all the solutions are quite different from the method of higher taxon delimitation (i.e. by drawing phenon lines) used in taxometrics.

Notwithstanding deviations that some cladists advocate and practise, the ideal is still to recognize as taxa monophyletic groups that are defined by synapomorphies. Such groups are monothetic, in strong contrast to taxa defined in taxometrics, which are polythetic. This is seen by many as a disadvantage, for it tends to accentuate single unique features at the expense of several nearly unique ones. Insistence upon all members of a taxon possessing any particular apomorphous character can cause problems. It is quite likely, for example, that if the angiosperms were examined sufficiently thoroughly no characters would be found to be both unique and universal. For such reasons this rule is now relaxed by some cladists.

The cladograms shown in Figs 2.16 and 2.17 are termed **rooted** trees or dendrograms, i.e. their evolutionary polarity is decided and hence an ancestral taxon (usually the hypothetical one included among the EUs) is pinpointed. However, as mentioned above, some workers now advocate not specifying the polarity of characters, so that the dendrogram obtained is not directional; this is known as an **unrooted** tree or **network** (Fig. 2.18). Networks can, if desired, become rooted by deciding *a posteriori* which end is the most primitive. Such decisions are based upon the overall nature of the taxa possessing their particular combinations of character-states (as in the method of Sporne described earlier), rather than upon *a priori* consideration of individual characters.

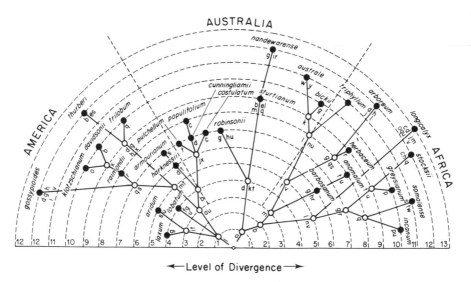

Fig. 2.17 Cladogram (Wagner tree) of 30 species of *Gossypium* (Malvaceae), modified from Fryxell[142].

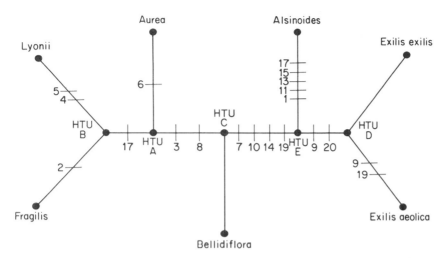

Fig. 2.18 Cladogram (unrooted tree or network) of seven taxa of *Pentachaeta* (Asteraccae), modified from Nelson and Van Horn.[303] Character-state transformations are represented by the numbers.

Cladistic methods basically view evolution as an ordered, divergent, step-wise transformation of characters from plesiomorphous to apomorphous states. It is known, however, that in evolution characters may sometimes evolve in a parallel or convergent fashion (Figs 2.6C and D), or may reverse direction so that an apomorphous character-state reverts to its plesiomorphous one (e.g. loss of stomata in some aquatic or chlorophyll-less angiosperms). Although convergence may cause the greatest distortion in cladograms, it is the most easily detected of these three phenomena. Character-states that arise by convergence are generally not logically correlated with other characters, and if a substantial number of characters is analysed those showing convergence are usually obvious. Parallelism and reversion, together termed *homoplasy*, are less easily recognized. The presence of homoplasy means that the correct cladogram might *not* be the one that appears to be the most parsimonious, and therefore methods based on parsimony are less than perfect in these cases. Reversals are allowed for in some parsimony methods, and there are ways of detecting parallelisms,[144,214,456] but the known frequency of homoplasy remains an important problem in parsimony methods.

Realization that one can rarely *for certain* deduce the phylogeny of a group solely from examination of extant organisms and known fossils has led some cladists to use cladistics (*pattern* or *transformed cladistics*) merely to unravel the pattern of variation rather than detect the true genealogy.[321,326] They believe that the pattern is nearly always close to the genealogy, and sometimes coincident with it, but there is no certainty that it is the same. This philosophy of transformation is strongly resisted by more traditional cladists.

Apart from the above parsimony methods, there are methods that utilize the concept of *character compatibility* and are known as *compatibility analysis* or *clique analysis*.[120,278] These have the advantage that they can detect and therefore omit homoplasy. As with parsimony methods, they can be carried out manually or via a computer program, and the characters used can be evolutionarily polarized or not to produce a rooted tree or network respectively. Groups of mutually compatible characters are termed *cliques*, and are considered to represent a group of characters in which homoplasy is absent. If we consider two characters (A and B) each with two states (A1 and A2, B1 and B2), four character-state combinations are possible—A1B1, A1B2, A2B1 and A2B2. Taking the direction of evolution to be A1 to A2 and B1 to B2, then if all four character-state combinations are found in nature there must have been at least one reversal (e.g. A2 to A1) or parallelism (e.g. A1 to A2 occurring twice). If so, A and B are incompatible, but if only two or three of the four combinations occur then A and B are compatible. Cliques are formed by comparing all pairs of characters and finding mutually compatible sets. From the data the largest clique is selected in order to produce the cladogram. The method is described in detail by Meacham,[278] but in essence it involves using one character (A) to separate the EUs into two groups (those with A1 and A2 respectively), and then further resolving each of the two groups by successively using further suitable characters. Finally, a rooted tree or a network (Fig. 2.19) is obtained according to whether or not a hypothetical ancestor was included among the EUs. There is argument as to whether the polarity of character transformations should be carried out *a priori* or *a posteriori*;[102,279,280] in the writer's view the latter is preferable as it is less prone to error.

Some authors consider that character compatibility is not a true cladistic method at all, whereas others believe it falls closer to the original methods of Hennig than do the parsimony methods of Wagner. In the writer's opinion there is little profit in such arguments. Any methods that give new insights into routes of evolution or new methods of classification are worthy of consideration, whatever their categorization and whatever they are called; they should all be tried and the one that best helps solve a given problem be utilized.

Cladistics has been adopted as a methodology far more readily and widely by zoologists than by botanists. There may be several reasons for this, but the botanists' caution is in fact to some extent justified by what seems to be a genuine difference in emphasis in the modes of evolution of at least higher plants and most animals. Hybridization is known to be common among higher plants and a major route in the evolution of new species (see Chapter 6). If the genealogy of a hybrid species is drawn out in the form of a cladogram it will, of course, show reticulations. Since cladistic methodology is based upon detecting phylogenetic branching by charting character transformations, it is normally unable to reveal reticulations. Actual points of reticulation are normally misinterpreted, or 'they obscure the underlying hierarchy'.[145] In fact the existence of hybrid EUs often causes the appearance of trichotomies or polychotomies in the cladogram; this is not a correct depiction of the actual situation. This major problem has been addressed on many occasions in recent years,[145,457,464] with very varied remedies similar only in that they are all unsatisfactory. For example, it has been suggested that hybrids should first be removed from the group of EUs under study, and then later be added to the completed cladogram of non-hybrid EUs. However, apart from obvious cases

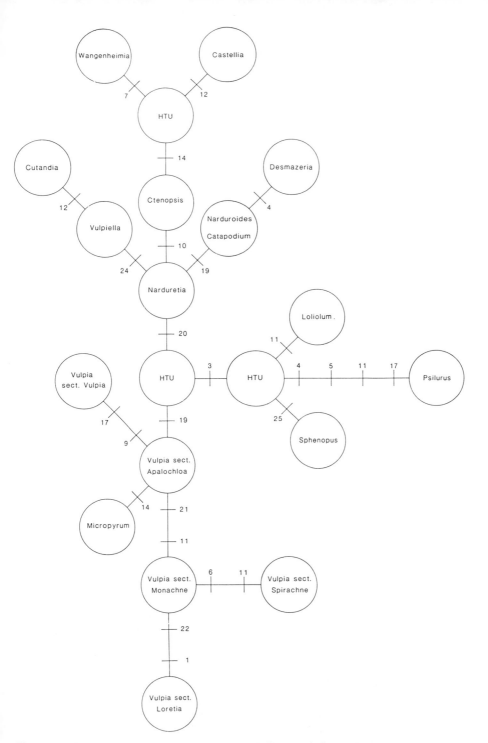

Fig. 2.19 Network constructed from compatibility analysis carried out according to Meacham's manual method by D. Shinn. The EUs are the genera listed in Figs. 2.11 and 2.14, plus two others and with *Vulpia* divided into its five sections. Character-state transformations are indicated by the numbers. The tree can be unequivocally rooted at *Vulpia* sect. *Loretia* by *a posteriori* reasoning. The tree agrees closely with the current classification; obvious suggestions are that *Psilurus* is relatively isolated, and that *Micropyrum* might be better amalgamated with *Vulpia* as a sixth section. Drawn by Mrs S. Ogden.

of neopolyploids, there is no certain way of detecting hybridogenous taxa, especially where the parents are now extinct. Because of this, others believe that hybrids should be retained in the primary analysis, despite the fact that interpreting them correctly is largely beyond current methodology. Hybridization remains a widespread and fundamental obstacle to the application of cladistics in many plant groups.

Cladists are all agreed that a prerequisite for cladistic analysis is a thorough knowledge of the group concerned. It is essential that the group and its constituent EUs be properly circumscribed, and that the characteristics to be used be understood sufficiently for homologies to be demonstrated and character transformation to be correctly assigned. Exactly the same prerequisite holds for taxometric analysis. The gaining of this knowledge and application is subjective. However objective the computational methods applied to the basic data might be, the results of analysis are not fully objective, since the use of different data will produce a different end result. Moreover, the interpretation of the cladograms in terms of a classification are also subjective; for example cladists argue whether convexity or monophyly should be the hallmark of a taxon. 'Objectivity is a myth.'[369]

The fact that successful cladistic and taxometric analyses are feasible only with well-studied groups indicates that the basic data used for these approaches, and indeed for 'narrative' methods, are the same. (Narrative methods are those which do not utilize an explicit, testable analytical approach, but rely heavily on taxonomic expertise and intuition; most of the work described under Phase 6 comes in this category). This offers the possibility that unless one or more of these approaches is seriously faulty, very similar or even identical dendrograms might result from whatever methodology is used. There is much evidence that this is often so. One example from several published is shown by a study of the monoraphid diatom family Achnanthaceae, from which cladograms and phenograms showed the same branching pattern.[251] The network shown in Fig. 2.19, and its rooted derivative, agree very closely with the current classification based on narrative methods, and phenograms of the same group are also extremely similar (Stace ined.). To many taxonomists it is not surprising that knowledge and ideas accumulated over many years of study can often produce results at least on a par with the most rigorous analytical methods.

If evolution proceeds at a constant rate and in a strictly divergent fashion, a correct phenogram will indeed appear exactly the same as a correct cladogram, and a highly competent taxonomist might subjectively produce the same pattern as well. However, if the above conditions do not hold, a phenogram and a cladogram will not look the same. In such cases the most important practical question to be posed is 'which is the *best* solution to adopt?' This has already been answered in Chapter 1 (p. 10). The firm stance taken by the writer is that predictivity is the criterion to be adopted, and the most predictive classifications should be considered the most natural one. It is not realistic or profitable to dictate *a priori* that cladistic or phenetic (or, indeed, narrative) methods might best achieve this. Such a view is today most often taken by cladists, but it is clear that cladistic classifications are not *always* the most predictive. If we consider a species (A) with several geographically separate subspecies (Aa, Ab . . . An), it could be that *one* of these subspecies (say Ab) gives rise, under extreme selective conditions imposed upon one or more of its subpopulations, to species B. Species B might well differ from species A by several apomorphous character-states, more than differentiate any of the constituent subspecies of species A. Hence phenetically species A and B are the most distinct entities, but cladistically species B and subspecies Ab are very close. In fact species A has become paraphyletic, but in this case the most predictive classification is obtained by recognizing it as a taxon. Exactly that conclusion was reached, for example, by Rahn[336] in his study of *Plantago*.

As was concluded at the end of the discussion of Phase 7 (taxometrics), cladistics is, in the writer's view, not likely to *replace* other disciplines, but to supplement them. *Often* its use will increase our understanding of a group and improve our classification. Paying heed to its methodology will *always* improve our approaches to data collection and manipulation and to the use of the results of taxonomic research in classification. But dogmatic insistence that the cladistic approach is always the best one, or the only valid one, will greatly hinder progress.

Section 2

Sources of Taxonomic Information

Introduction

Any data which show differences from species to species are of taxonomic significance, and thus constitute part of the information or evidence which may be used by taxonomists. Moreover, as has already been stressed, it is desirable that a wide range of information from diverse sources should be utilized in order to obtain the best sort of natural classification. It has also been mentioned that taxonomy is a science without data of its own, utilizing the results of investigations in all other branches of biology. This is true in theory, but in practice taxonomists are often not able to produce classifications by the manipulation of data produced by others, because of lack of sufficient data. The results of chemical, genetical, anatomical, cytological and other investigations are relatively seldom expressed in systematic terms, i.e. they are not usually from comparative studies of a wide range of organisms, but much more often involve one or few taxa as examples of a great many.

Hence, for the most part, taxonomists have to gather their data themselves, and in fact most taxonomists spend most of their time engaged in this essentially non-taxonomic operation. Bearing in mind the enormous number of species which they have to deal with, and the virtually endless array of possible taxonomic characters, it is hardly surprising that taxonomists have, of necessity, been very selective in the characters which they have chosen to study. Characters selected are often those which are most easily observed, and those which show promise as being reliable and discriminating in taxon delimitation. This is not altogether undesirable, because it is most convenient if taxa are delimited by obvious features rather than by cryptic ones, but it is one of the major causes of the considerable degree of subjectivity in taxonomy, and this remains the most frequent object of criticism by non-taxonomists.

Many less obvious sorts of characters have been utilized by taxonomists only in relatively modern times, both because taxonomists did not previously have the time, or see the need, to investigate them, and because in many cases the necessary apparatus, methods and expertise have only recently become available, at least to taxonomists. By chance, such 'new' techniques often produce remarkably valuable data of great use in classification, and this has given rise to the notion, still held by many, that new techniques uncover more reliable or fundamental characters than old ones. Features such as chromosome number and morphology, pollen structure, ability or inability to interbreed, stomatal architecture, occurrence of secondary metabolites, protein amino-acid sequences and so on have all at various times been held up as 'special cases' in this way, and most of them are still able to attract government grants far more readily than is the use of 'traditional' techniques. Actually none of these novelties is any more (or less) taxonomically valuable than are characters of gross morphology, because instances of their spectacular success in classification are countered by a greater number of cases where they are of little or no value. Moreover, in a group of plants in which a particular range of features has not been investigated, it is quite impossible to predict accurately whether those features will prove of great, moderate or low taxonomic significance. This theme will repeatedly emerge from the next few chapters.

For convenience, taxonomic information is discussed under separate headings, although this tends to produce an unwelcome compartmentalization of ideas when the unity and equality of different taxonomic characters would be better emphasized. Davis and Heywood[98] divided their discussion into morphology and anatomy, cytology, and phytochemistry, whereas Benson[31] used the headings herbarium studies, field observations, microscopic morphology, palaeobotany and biogeography, chemistry and ecology, and cytogenetics. Sneath and Sokal[393] considered four main types of characters: morphological (including anatomical), physiological and chemical, behavioural, and ecological and distributional. The chapter headings in this book are different from any of the above, but it must be stressed that they have been delimited simply for convenience and are not necessarily considered better than other classifications in all contexts.

3
Structural information

In this chapter the subjects usually known as morphology and anatomy are covered; molecular structure will be discussed in Chapter 4, and chromosomal structure in Chapter 5.

The view is taken here that there is no logic in the artificial separation of discussions on different aspects of structural botany for taxonomic purposes. Hence the following topics are considered together: morphology and anatomy; reproductive and vegetative structure; structure of modern and of fossil plants (neobotany and palaeobotany); and mature and developmental structure. Various specialized fields, such as palynology, embryology and morphogenesis, therefore, belong here too. It is considered, in fact, that it is often the separation of many of these structural aspects in the past that has led to biased opinions on the relative taxonomic values of the evidence obtained by specialists. There is, *a priori*, no reason to believe that the position of the ovary is more important than that of stomata, that the anatomy of timber is more significant than the morphology of the stem, or that the developmental stages in the growth of hairs are more useful than their mature structure. Claims of this nature have often been made, sometimes with, but more often without, justification.

Plant taxonomy was founded upon the characters covered in this chapter, and today they still provide the bulk of taxonomic evidence. No attempt is made here (or could be made) to cover all the types of structural information. Instead, general points are illustrated by reference to a wide range of examples.

Reproductive and vegetative characters

It is well known that floral characters have been, and still are, those most used in the classification of angiosperms. At one time elementary botanical courses were laden with innumerable, often complicated (and not infrequently redundant) terms which had been invented to describe a vast array of floral variation (Fig. 3.1). Every conceivable facet of the variation of the inflorescence, bracts, receptacle, hypanthium, calyx, corolla, nectaries, disk, stamens, carpels and ovules has been found by taxonomists to be valuable in the classification of one or other group of angiosperms. In the diagnostic keys to taxa in Floras, floral characters are those most used, and they figure equally prominently in the descriptions of the taxa which follow the keys. Examples could be given of the use of such features at all taxonomic levels. This reliance

PERIANTH and ANDROECIAL POSITION

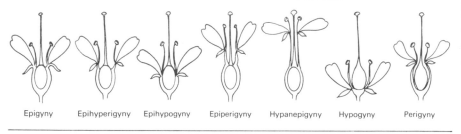

| Epigyny | Epihyperigyny | Epihypogyny | Epiperigyny | Hypanepigyny | Hypogyny | Perigyny |

PLACENTAL POSITION

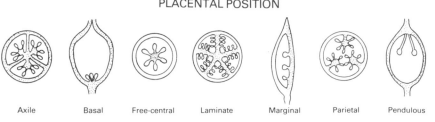

| Axile | Basal | Free-central | Laminate | Marginal | Parietal | Pendulous |

INFLORESCENCE TYPES

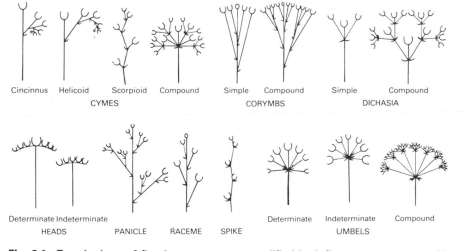

Cincinnus   Helicoid   Scorpioid   Compound   Simple   Compound   Simple   Compound
                    CYMES   CORYMBS   DICHASIA

Determinate Indeterminate   Determinate   Indeterminate   Compound
     HEADS   PANICLE   RACEME   SPIKE   UMBELS

Fig. 3.1 Terminology of floral structure as exemplified by inflorescence types, and by perianth, androecium and placenta position, reproduced from Radford *et al.*[335]

on the flower is remarkable when one considers that for most of the time the majority of angiosperms lack any flowers at all!

From this vast pool of experience certain generalizations can be made with regard to the characters found to be most useful in certain taxa, or the level of conservativeness of certain characters. For example, in the Ranunculaceae the bracts, sepals and petals are particularly important, in the Asteraceae and

Poaceae the inflorescence, bracts and general flower-type, in the Fabaceae the stamens and carpels, in the Scrophulariaceae and Lamiaceae the corolla and stamens, and so on. There are, however, very large numbers of exceptions to these generalizations, and too much reliance on the latter would result in important errors or oversights.

Although the broad pattern of the structure of fruits and seeds is in general a less reliable taxonomic criterion (superficially similar fruits or seeds sometimes being found in only remotely related groups), they are often of supreme importance. In such large families as the Brassicaceae and Apiaceae, for example, the fruits provide the most useful characters of all, and they are scarcely less important in the Rosaceae. In such families the number of special terms used to describe the variation of the fruits is, as expected, high. Beginners in field botany soon learn that care must be taken when collecting members of the Brassicaceae, for example, to include ripe fruits with the specimen in order to guarantee identification by means of a Flora.

Similarly, seeds are of particular significance in families such as the Caryophyllaceae. In this family it is not simply that seeds are useful in delimiting tribes and genera, etc., but that, in many cases, seed-structure produces taxonomic criteria at all levels of the hierarchy. The genera *Silene*, *Dianthus*, *Petrorhagia*, *Stellaria*, *Spergularia* and *Spergula* illustrate this well. Whereas seed-structure usually differs interspecifically in these genera, in certain cases it varies at an infraspecific level (e.g. *Spergula arvensis*) or even within an individual (e.g. *Spergularia media*). The four European subspecies of *Montia fontana* (Portulacaceae) are distinguished almost solely on seed morphology[460] (Fig. 3.2). Seed characters are particularly investigated by botanists interested in fossil (especially Quaternary) floras, as the seeds are often the only parts remaining. The use of seed characters, from gross morphological to subcellular levels, is today rapidly increasing.[17]

In lower plants flowers are lacking, and taxonomists have always had to rely on other characters. In algae and bryophytes, particularly, reproductive organs are often rather rare or shortlived, and most classifications have in the past been based upon vegetative structure. By diligent searching and careful recording over the past two centuries we have now amassed a great deal of information on the reproductive features of these groups, and have therefore obtained many new taxonomic criteria. In some cases these data have thrown a different light on the relationships of lower plants, but in the main they have caused taxonomic adjustments rather than revolutions. In bryophytes, for example, the sex organs are very conservative (see Chapter 8), and in the case of the liverworts (hepatics) this applies to the sporophytes as well. In mosses the sporophytes are less conservative and have provided many extra characters which have led to some notable changes in the classification. Nevertheless, the identification of bryophytes is still carried out using mainly vegetative characters, for obvious reasons. In the algae the reproductive organs provide many taxonomic characters, but these largely parallel the delimitation of taxa indicated by the vegetative structure and chemistry. In lower plants as a whole, in fact, the distinction between vegetative and reproductive structures is blurred; the sporophyte of a bryophyte, or the gametophyte of a pteridophyte, for example, could be regarded as either vegetative or reproductive.

0·25 mm

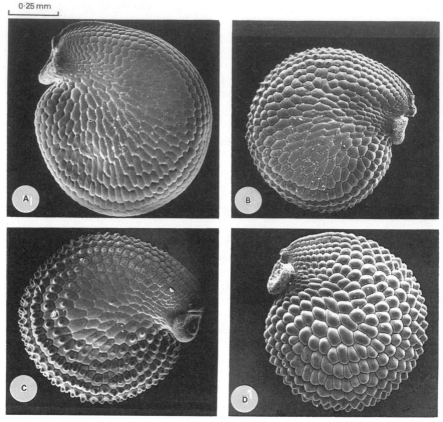

Fig. 3.2 Scanning electron micrographs of seeds of the four subspecies of *Montia fontana*: **A**, subsp. *fontana*; **B**, subsp. *variabilis*; **C**, subsp. *amporitana*; **D**, subsp. *chondrosperma*. Photographs by A. N. Scott.

In the higher plants vegetative characters are often looked upon as risky evidence, because there are many cases where superficially similar morphological features are found in quite unrelated plants. Similar growth-habits (e.g. cactus-like plants, broom-like plants, trees) or leaf-shapes (e.g. palmately-lobed, compound pinnate) are good examples. The similarity in the leaf-shape of various maples (*Acer*) and planes (*Platanus*), which belong to very remote families, are well known; in Britain the name sycamore is applied to a species of maple, whereas in America it refers to a plane-tree, and the sycamore of the bible is yet another plant (a fig) with a similar leaf-shape. Thus vegetative characters have been used rather reluctantly, and particularly in groups where reproductive characters are unhelpful in classification. The grasses (Poaceae) are in this category. In the elms (*Ulmus*) the flower and fruit are of little taxonomic value, and the shape of the leaves provides the most important taxonomic features. In oaks (*Quercus*) and birches (*Betula*) the leaf characters are again the most important ones. Within the family Ranunculaceae the genera *Aquilegia* and *Thalictrum* differ by many important characters, especially the entirely different flower structure and fruit-types (follicle and achene respectively). The leaves, however, are very similar, as evidenced by a species of *Aquilegia* named *A. thalictrifolia*, and one of *Thalictrum* named *T. aquilegifolium*. Since the family Ranunculaceae was formerly divided into two tribes or subfamilies on the basis of the two fruit-types, these two genera fell into different groups (even, in Hutchinson's scheme, into separate families). However, study of the chromosomes of the Ranunculaceae[167] showed that *Aquilegia* and *Thalictrum*, along with a few others, differ from most Ranunculaceae in having small (rather than large) chromosomes and chromosome numbers based on multiples of seven (rather than eight). In this case, the vegetative characters are therefore considered to be more indicative of relationship than the reproductive ones, and these two genera are now segregated into a separate tribe.

The phenomenon of the incongruence of classifications using vegetative as opposed to reproductive features in both the fossil pteridosperms and the living *Ginkgo* has already been mentioned. Many other examples could be given, for example the moss *Andreaea*, which is similar to the true mosses vegetatively but to *Sphagnum* on capsule characters. The existence of such situations merely serves to emphasize that the best classifications are produced by the use of both reproductive and vegetative characters.

Morphological and anatomical characters

The use of anatomical characters in taxonomy has been confined to the last 100 years or so, since high-power microscopes became commonly available, and even today microscopic structure is often regarded as an adjunct to macroscopic features rather than as a standard source of data. This is not surprising, for it is clearly easier to determine whether the petals are free or fused than it is to decide whether the vessels have simple or multiple perforation plates. There has, however, been a remarkable revolution in the past 40 or so years in the investigation of vascular plant anatomy and its use in classification; many new studies have been published, as well as a number of good reviews.[20,22,23,86,100,124,290,363,439] It is now generally realized that anatomical characters are just as valuable as morphological ones, and must not be neglected. In fact there is a body of opinion that holds the former to be more revealing than the latter. The similar leaves of *Acer* and *Platanus*, for example, have different anatomical features, and it is indeed often possible to interpret vestigial or modified parts of flowers by studying their vascularization, on the grounds that the internal parts of a plant are less affected by the environment (both genetically and phenetically) than the superficial parts. In the Chenopodiaceae the anomalous, distinctive stem anatomy is common to a series of genera of very different growth-habits; in *Euphorbia* all the species are characterized by the presence of latex-vessels, whether the plants be cactus-like, thorny shrubs, or leafy herbs. On the other hand, the absence of conducting tissues in water plants, the scattered taxonomic occurrence of many cell-types and crystal forms, and the presence of bicollateral vascular bundles in climbing plants are instances which bring a note of warning to too sweeping a generalization.

In very small plants, particularly algae and bryophytes, anatomical characters have been used much more extensively, because of the lack of more obvious features. In bryophyte Floras, for example, anatomical characters (especially leaf cell-shape) form standard parts of the descriptions and keys, and the bryologist is resigned to the fact that many taxa cannot be identified for certainty without the use of a microscope. In the case of unicellular algae this is invariably the case, and of course in such plants morphology *is* anatomy. The realization that morphology and anatomy are inseparable aspects of structure has led to the use of alternative pairs of terms such as **macromorphology** and **micromorphology**, and **exomorphology** (external) and **endomorphology** (internal). The former of these two pairs is no less subjective than morphology and anatomy, but has become useful in the context of the scanning electron microscope (S.E.M.), the term micromorphology generally being applied to electron microscopic features. The term **ultrastructure** is more usually used for structure seen with the transmission electron microscope (T.E.M.), though any distinction is now obsolete.

Nowadays virtually every anatomical aspect of plants has been studied by taxonomists, and the quantity of information accumulated is enormous. Retrieval and use of this information are made easier by various bibliographic compilations. Most notable of these is *Anatomy of the Dicotyledons* (1950)[291] (second edition 1979 onwards)[292] and *Anatomy of the Monocotyledons* (1960[289] onwards), of which seven volumes have so far appeared under the editorship of C. R. Metcalfe at the Royal Botanic Gardens, Kew. It must be noted, however, that these volumes include only *vegetative* anatomy (not reproductive anatomy), and that no such compilations exist for the pteridophytes and gymnosperms. It is difficult to envisage these gaps being filled this century, although a number of special systematic reviews exist, for example Florin's work on the epidermis and cuticle of gymnosperms (1931, 1933),[136,137] *The Seeds of Dicotyledons* (1976),[73] *World Pollen and Spore Flora* (1973 onwards),[311] *The Northwest European Pollen Flora* (1976 onwards)[334] and the Forest Products Research Bulletins *Identification of Hardwoods* (1960, 1961)[11,48] and *Identification of Softwoods* (1948).[324] In the bryophytes and algae, as noted before, standard taxonomic works are largely concerned with anatomical data.

More numerous are various non-systematic surveys of the taxonomic value of particular organs or groups of characters, or glossaries attempting to redefine precisely the terms recommended for use. Good examples are found relating to pollen,[38,118,125,458] cuticle and epidermis,[101,409,478] stomata,[222] leaf venation,[101,207,208,287] leaf shape (Fig. 3.3),[122,123] ptyxis[84] and trichomes.[322,337,455]

Electron microscopy has of course provided an insight into structure at a much lower order of magnitude than conventional microscopy. The S.E.M., at least at its more usual (lower) levels of resolution, has uncovered relatively few *new* characters, but has enabled the rapid, recordable and comparative study of a great many micromorphological features, so that these have become realistic and practicable as standard taxonomic characters. Spores (including pollen grains), leaf surfaces (especially stomatal architecture), seed (Fig. 3.2) and fruit surfaces, and microscopic plants (notably diatoms, Fig. 3.4A) are among the subjects to have particularly benefited so far. The use of scanning electron micrographs is now commonplace in the taxonomic literature.[202]

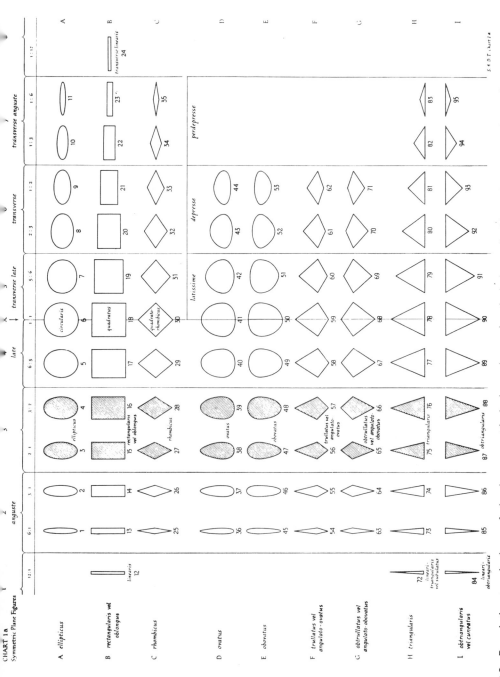

Fig. 3.3 Descriptive terminology of simple symmetrical plane shapes (particularly applicable to leaves, petals, etc.) as suggested by a committee of the Systematics Association, taken from Exell.[123]

A

10 μm

B

1 μm

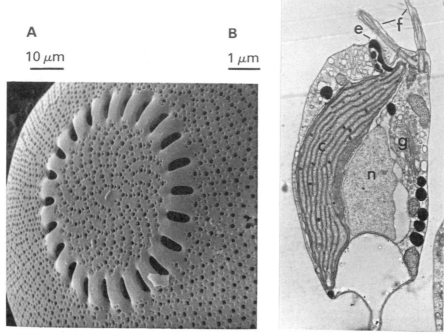

Fig. 3.4 Electron micrographs of algae: **A,** Scanning electron micrograph of part of cell-wall of *Brightwellia coronata*, a fossil marine diatom (Bacillariophyta) from New Zealand, taken from Ross and Sims.[366] **B,** Transmission electron micrograph of longitudinal section of a *Uroglena* species (Chrysophyta), taken from Hibberd[206] (c = chloroplast, g = golgi body, n = nucleus, e = eye-spot, f = two flagella).

By contrast, the T.E.M. is a less handy tool but has no realistic alternative in the very high resolution study of internal structure. The taxonomic value of its discoveries so far shows a remarkable contrast between higher and lower plants. In the flowering plants there are very few important diagnostic features at the high taxonomic levels.[22,23] Dilated cisternae of the endoplasmic reticulum of several cell-types occur only in families of the order Capparales (notably Brassicaceae),[21,242] where they seem to be functionally related to the presence of glucosinolates, a group of Capparales-specific secondary plant products, and of the equally restricted myrosin-cells, which contain the enzyme myrosinase known to hydrolyse glucosinolates. The various types of plastids found in the sieve-tubes of the phloem throughout the angiosperms, defined primarily by the presence and form of protein-accumulations, are highly diagnostic and can be used to characterize taxa such as the monocotyledons and the order Caryophyllales (Fig. 3.5). Other features, such as nuclear protein-crystals and seed protein-bodies, seem promising characters for the future.[23] At the lower levels of the hierarchy ultrastructural evidence can provide information either strengthening conclusions based on other data or indicating new relationships, for example the segregation of ecotypes of *Xanthium* based on plastid organization.[3] In the use of pollen characters,

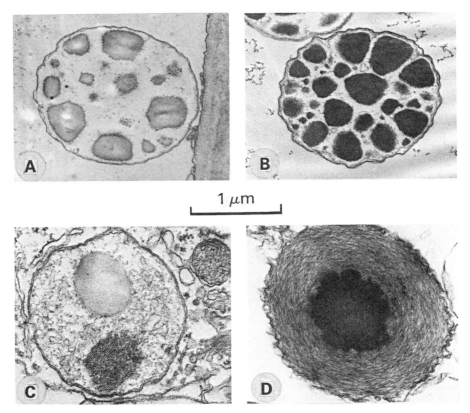

Fig. 3.5 Transmission electron micrographs of four angiosperm sieve-tube plastids: **A,** *Ocotea* (Lauraceae); **B,** *Dracaena* (Agavaceae); **C,** *Hernandia* (Hernandiaceae); **D,** *Petiveria* (Phytolaccaceae). Taken from Behnke.[20]

interpreting surface features observed with the S.E.M. has been greatly facilitated by the study of thin sections with the T.E.M.[320]

In the lower plants, however, ultrastructural details, along with chemical data, have revolutionized our taxonomic concepts. To a large degree the ultrastructural details were uncovered after the chemical ones, and the former served to confirm taxonomic conclusions based on the latter, but new ideas also emerged. Perhaps most remarkable was the discovery that the blue-green algae, in their cell-wall structure, ribosome characteristics and lack of endomembranes (among other features), are more closely related to bacteria than to algae; nowadays these two groups are usually placed in a separate kingdom, the Procaryota. Within the true algae many ultrastructural characters, notably of the plastids, have served to strengthen and clarify our views on the fundamental differences between the major divisions, for example Rhodophyta, Phaeophyta, Chrysophyta and Chlorophyta (Fig. 3.4B). New work on the ultrastructure of the Charales (stoneworts),[338,425] coupled with light microscopy and biochemical studies, has shown that this group is less closely related to most of the Chlorophyta (green algae) than hitherto believed. However, certain groups of the latter, for example Zygnematales and Coleochaetales, also possess the distinctive Charalian characteristics, and the separation at the level of class of a newly defined (broadened) Charophyceae from the Chlorophyceae seems taxonomically advisable. Further, it has been found that the cellulosic, plant-like fungi (Oomycetes) possess golgi-bodies like those in green plants, whereas the other (chitinous) fungi lack such organelles, thus strengthening the belief that the former (but not the latter) are colourless plants.

Particularly valuable taxonomic evidence has been obtained from the study of pollen, wood, leaf epidermis and cuticle, trichomes and stomata. Some of these anatomical features are so diagnostic that they are now commonly used in routine identification, rather than being confined to a use in problems of phylogeny or classification, or in the identification of fragments of plants. Hence trichome characters are important in the determination of members of the Brassicaceae and the genera *Combretum*, *Rhododendron* and *Epilobium*, and pollen characters in the identification of species of *Callitriche*. But anatomical features are of particular value to scientists who need to identify small scraps of plant material, for example pharmacognosists (in drugs), forensic experts (as clues) and animal dieticians (in gut and faeces).

Three particular examples may be used to illustrate the value of anatomical characters. In the tropical family Combretaceae, Stace[403,408] found that trichome anatomy is of immense significance in classification at all levels, from the circumscription of the family down to the separation of species and even varieties. In particular, it has led to an improved tribal classification within the family and an improved subgeneric and sectional classification within the largest genus *Combretum* (Fig. 3.6).

Among the many taxonomically important features of stomata, the arrangement of the surrounding epidermal cells (termed **subsidiary cells** if they are distinct from the normal epidermal cells) is the most valuable. Thirty-five distinct patterns have been found in the vascular plants as a whole, including a number known only in the pteridophytes[101,222,409] (Fig. 3.7). The occurrence of these types is often valuable at the higher taxonomic levels. Thus in the Acanthaceae the stomata are diacytic, whereas in the closely related Scrophulariaceae they are anomocytic; within the family Combretaceae the stomata are paracytic in the subfamily Strephonematoideae and anomocytic in the subfamily Combretoideae, apart from in the tribe Laguncularieae, where they are cyclocytic.[403] Stomatal characteristics are not, however, always so reliable, and the different types have obviously each originated on many different occasions and they often exist together on one plant. In the genera *Streptocarpus* and *Saintpaulia* of the Gesneriaceae, for example, the cotyledons bear anomocytic stomata and the mature organs anisocytic stomata,[370] and in *Lippia nodiflora* (Verbenaceae) anomocytic, anisocytic, diacytic and paracytic stomata occur on the same leaf.[317]

In the family Poaceae (grasses) the flowers are greatly reduced and provide a rather limited number of taxonomic characters, so that more attention has been paid to vegetative anatomy and to cytology. Fortunately the leaf epidermis (Fig. 3.8) and the leaf cross-sectional anatomy provide extensive taxonomic data, and the literature on this subject is now vast.[116,117,289] Characters such as the disposition of sclerenchyma, the arrangement and

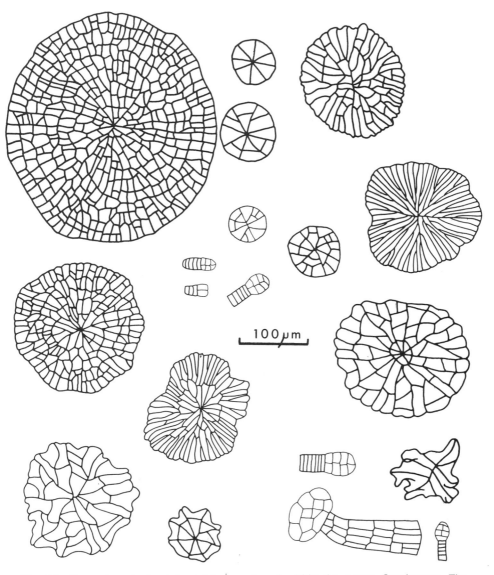

Fig. 3.6 Examples of glandular trichomes found within the genus *Combretum*. The trichomes illustrated are taken from species from different sections of the genus *Combretum*. Six stalked glands are shown in side view; the others are peltate glands shown in surface view.

form of the vascular bundles, the differentiation of epidermal long-cells and short-cells, and the form and distribution of silica-bodies and various types of trichomes and papillae have played a big part in the modern re-classification of the family at all levels from subfamily to species, and all thorough taxonomic studies of grasses now pay considerable attention to them.

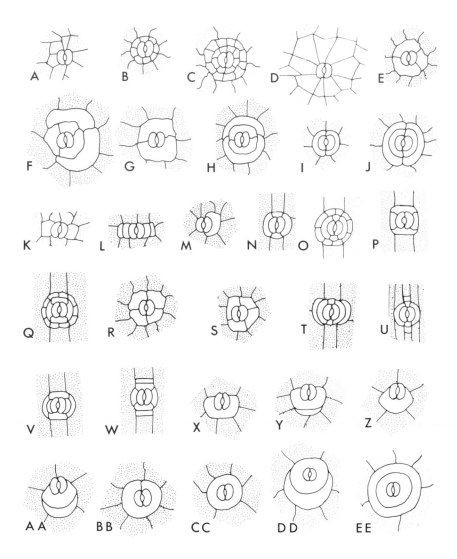

Fig. 3.7 Thirty-one types of arrangement of subsidiary cells in the mature stomatal complex of vascular plants, adapted from Dilcher:[101] **A**, anomocytic; **B**, cyclocytic; **C**, amphicyclocytic; **D**, actinocytic; **E**, anisocytic; **F**, amphianisocytic; **G**, diacytic; **H**, amphidiacytic; **I**, paracytic; **J**, amphiparacytic; **K**, brachyparacytic; **L**, amphibrachyparacytic; **M**, hemiparacytic; **N**, paratetracytic; **O**, amphiparatetracytic; **P**, brachyparatetracytic; **Q**, amphibrachyparatetracytic; **R**, staurocytic; **S**, anomotetracytic; **T**, parahexacytic-monopolar; **U**, parahexacytic-dipolar; **V**, brachyparahexacytic-monopolar; **W**, brachyparahexacytic-dipolar; **X**, polocytic; **Y**, copolocytic; **Z**, axillocytic; **AA**, coaxillocytic; **BB**, desmocytic; **CC**, pericytic; **DD**, copericytic; **EE**, amphipericytic. Four other types now recognized were not known to Dilcher.

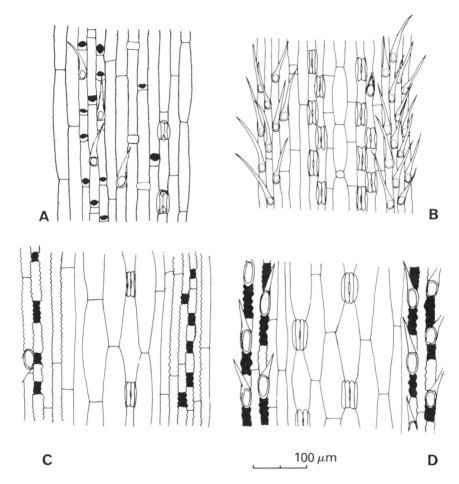

Fig. 3.8 Epidermises of the grasses *Vulpia alopecuros* (**A**, abaxial; **B**, adaxial) and *Vulpiella tenuis* (**C**, abaxial; **D**, adaxial), taken from Cotton and Stace.[75] Silica-cells are shown in black. Until recently the single species of *Vulpiella* was included as a species of *Vulpia*, but it differs markedly in microscopic characters such as those illustrated.

Mature and developmental characters

It is frequently necessary to investigate the development of an organism, organ or character in order to allow its proper interpretation and hence its full use in taxonomy. Moreover, during such investigations further taxonomic characters of a developmental nature are often discovered. This principle can be illustrated at all levels.

It has long been recognized that in the vascular plants the different mature stomatal types often arise by quite different modes of development. Florin[136,137] recognized two main modes in the gymnosperms, based upon whether or not the guard-cells and subsidiary cells originate from the same mother-cell (**syndetocheile** and **haplocheile** respectively). In the angiosperms

Pant[316] recognized the same two categories (termed **mesogenous** and **perigenous** respectively), and a third (**mesoperigenous**) where some adjacent cells were and some were not derived from guard-cell initials. He further subdivided these three main categories into ten categories in all, and more recent authors have increased this total considerably.[222,409] It must be emphasized that there is no strict correlation between the mature stomatal type and the mode of development; in particular the presence of subsidiary cells does not necessarily indicate a mesogenous development (and their absence a perigenous one), as Florin supposed. In some schemes[222] the two methods of classification are combined, whereby each mode of development is described by a term with a prefix denoting the mature type and a suffix denoting one of the three main modes of development—hence tetra-perigenous, dia-mesogenous, etc. The study of stomatal development and the application of these descriptive terms, which are now equally applied to all vascular plants, often elicits more useful taxonomic characters than the use of mature characteristics alone, especially when distinctive subsidiary cells develop at a late stage in a perigenous fashion, or when a mesogenous development does not give rise to distinguishable subsidiary cells. However, developmental studies do not always give extra data. In two of the cases mentioned above, for example, the anomocytic and anisocytic stomata of *Streptocarpus* develop perigenously and mesogenously respectively, and the leaves of *Lippia* bear stomata which develop in four different patterns.

Similar examples to the above could be drawn from studies in trichome development and elsewhere. The comparison of leaves and other organs is often meaningful only if carried out at the same stage of development, especially in plants showing heterophylly, such as *Vicia sativa, Ranunculus aquatilis* and *Populus tremula*. Melville,[286] in his study of the genus *Ulmus*, found it necessary to compare **leaf spectra** rather than individual leaves. In families such as the Brassicaceae and Apiaceae it is often necessary to examine pre-flowering plants in order to observe the diagnostically important rosette leaves, which may wither before flowering. Cotyledon characters are likewise important in the families Combretaceae and Gesneriaceae, and in the genera *Phaseolus* and *Ranunculus*, among others.

The life-cycle of plants is not only different from that of animals in that (with the possible exception of a few algae such as diatoms and the Fucales) gametes are produced by the haploid gametophyte generation, not directly by meiosis, but it is also far more varied. This variation is of vital significance in the classification of the major groups of plants, for example in the bryophytes, pteridophytes and spermatophytes, and particularly so within the algae. Until the life cycles of many of the lesser-known algae were uncovered different generations of the same species were sometimes given separate names and placed in separate parts of the classification system; the same is still true of some fungi and microscopic algae.

Features of sporogenesis, gametogenesis, fertilization and embryogenesis in the flowering plants have proved particularly rich in new taxonomic characters. Quite recently the number of nuclei in the pollen at the time of dispersal (two or three according to the precocity of division of the generative nucleus) has been used by taxonomists;[51] it appears that the binucleate state is the primitive one and the trinucleate state has arisen from it in many different groups. Moreover, the number of nuclei is correlated with various physiological and genetical features, such as the type of photosynthetic pathway and the mechanism of self-incompatibility. The peculiar mode of development of the pollen grains of the Cyperaceae, whereby three of the four nuclear products of meiosis abort and only one pollen grain (***pseudo-monad***) is formed from each pollen mother-cell, is approached in the Juncaceae, where pollen grain formation is delayed until the pollen mother-cell has eight nuclei and the pollen is shed in tight tetrads, thus strengthening the idea that these two families are closely related. Double fertilization, the participation of both male nuclei in sexual fusion to form a true endosperm as well as an embryo, was once thought to be absolutely diagnostic of the angiosperms, i.e. occur in all angiosperms but in no other plants,[282] but it is now known that it is absent from all or most Orchidaceae and Podostemonaceae[488] and perhaps *Acorus*. Characters of the embryology of angiosperms have long been utilized by taxonomists.[90,97,235,313] The known variation in the form of the embryo sac is shown in Fig. 3.9.

Among the bryophytes and pteridophytes many parallel instances could be cited, for example the separation of the Anthocerotopsida from the Hepaticopsida by the origin of sporogenous tissue in the capsule from the amphithecium in the former but from the endothecium in the latter group, and by the antheridia arising from a subsurface cell in the former but from a superficial cell in the latter group.

Value of characters

As emphasized previously, any particular character varies enormously from group to group in its taxonomic value, and it is quite impossible to predict this value in a group in which that character has not been previously investigated. Stamen number, for example, can be a familial, generic or specific character, or vary greatly within one taxon. The same is true of dioecism, equally so in the bryophytes and flowering plants. Palynologists are well aware of the uniformity of pollen morphology within the Poaceae; were this not so pollen analysis would have revealed a far greater range of palaeo-ecological data than has been the case. Yet in families such as the Acanthaceae, pollen morphology is one of the most important sorts of taxonomic evidence; the same is true of the spores of, for example, the hepatic genus *Fossombronia*.[63]

In some instances very obvious characters are overlooked because they are usually quite invariable. Nevertheless the absence of vessels and the presence of unclosed carpels in certain allegedly primitive dicotyledons, and the presence of only one (or three or more) cotyledons in a wide range of dicotyledons, and of more than one cotyledon in several monocotyledons, emphasize that *all* characters are worthy of examination. Taxonomists are, in general, aware of the need to examine all sources of potential evidence, and it is less necessary nowadays to emphasize groups of neglected characters than was the case in 1963.[98] It is this awareness that has led taxonomists to investigate the distribution of obscure structures (e.g. dilated cisternae in root-cap cells) or organs often not preserved on herbarium sheets (e.g. underground parts of *Crocus* and *Oenanthe*, or tree 'architecture').[175]

Too much reliance, as indicators of taxonomic relationship (phenetic or

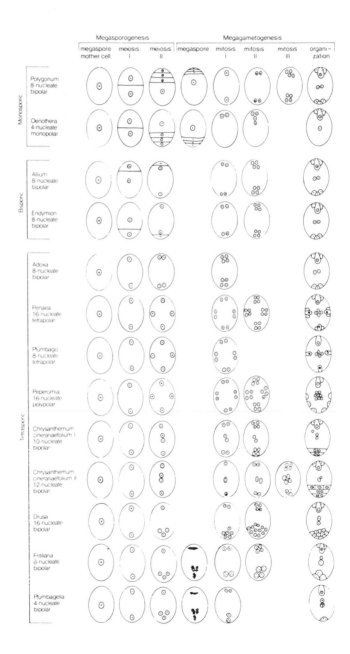

Fig. 3.9 Scheme illustrating the main modes of development and mature structure of embryo sacs in angiosperms, taken from Willemse and Van Went[497].

phylogenetic), should not be placed on characters which have been shown to have frequently resulted from evolutionary convergence. The arboreal or annual habit, dioecism, succulence, fleshy or winged fruits and seeds, and an isomorphic alternation of generations (or extreme reduction of the sporophyte or gametophyte) are obvious examples. Hamann[176] has shown that there are many convergences among the characters of angiosperm embryology, as has Jurgens[244] in leaf anatomy of unrelated succulents. Sometimes one finds a good correlation between two or more such characters, for example the absence of lodicules correlated with protogyny in the grasses *Anthoxanthum* and *Alopecurus*, and the reduction in petal size, nectaries and stamen number in self-pollinating Brassicaceae such as *Cardamine hirsuta* and *Arabidopsis thaliana*. Over-use of such features has led to the erection of notoriously polyphyletic taxa such as the Monochlamydeae (Apetalae) of Bentham and Hooker, the Amentiferae (catkin-bearing plants),[437] or the moss group Nematodonteae.

4
Chemical information

Plant chemotaxonomy (*chemosystematics*, *chemical plant taxonomy* or *systematics*, or simply *phytochemistry*) is one of the more fashionable and rapidly expanding areas of plant taxonomy and seeks to utilize chemical information to improve the classification of plants. It has arisen as a major science in only the past 30 or so years and, perhaps because of that, has two common fallacies associated with it. These are that chemotaxonomy is a recent innovation, and that it in some way deals with more fundamental information than does, say, morphology or cytology, so that the data which it produces are more important in classification.

Origin and nature of chemotaxonomy

In fact chemotaxonomy has many diverse and indeed ancient origins. Perhaps foremost comes the search by herbalists, and latterly by pharmacists, for drugs, which has involved the accumulation of information on the chemical content of a very wide range of plants. The statement that similar sorts of plants have, in general, similar medicinal properties, i.e. contain similar chemicals, is at least 300 years old, and the concept probably dates back thousands of years. Today wild plants are still used for pharmaceutical purposes and new sources of drugs are still being sought. Even before the use of drugs, ancient man sought plants as food and must have realized that similar species often possessed common properties. Thus, by trial and error, he discovered the high protein-content of seeds of the Poaceae (grasses) and Fabaceae (legumes).

A second major ancient origin of chemotaxonomy has been through the fields of morphology and anatomy. For example, colour can be regarded as either morphological or chemical, and different forms of crystals as either anatomical or chemical. It is of course now well known that any particular colour can be based upon the presence of any one or a combination of a great many different molecules, while crystals and other inclusions not only vary in their chemistry (calcium oxalate, calcium carbonate, starch, silica, etc.) but also in their physical structure. The different forms in which starch, silica and calcium oxalate are to be found in plant cells have been particularly valuable as taxonomic evidence. For example, 20 types of silica-body can be recognized in the grasses,[289] and needle-like forms (raphides) of calcium oxalate are of very limited occurrence in dicotyledons (notably Rubiaceae and Onagraceae among British representatives). About 14 types of starch grains can be recognized in the angiosperms and used to delimit taxa at all levels.[87]

The taste and smell of plants are not only detected and used as indicative of their food, drug or cosmetic value by humans, but must contribute to the selectivity shown by many animals and pathogens. In recent years much attention has been focused on the **co-evolution** of various plants and animals,[181] particularly concerning the preferences shown by animals in their use of plants for food. This phenomenon is perhaps best exhibited by insects, species of which are often confined to a single species of food-plant. In many cases related insects feed on related plants, for example the strong preference shown by the larvae of the butterfly family Danaidae for the Asclepiadaceae. Other examples concern browsing herbivorous mammals[56] or molluscs,[237] and the behaviour of pollinating insects.[14,245,283,332] Pathogenic fungi such as the rusts similarly show affinity for related host-plants, and have been used to investigate taxonomic problems in, for example, pteridophytes and the Liliaceae–Amaryllidaceae group of families.[26,284,371] Gardeners, too, have known for centuries that it is possible to graft only closely related taxa, for example pear and apple, *Laburnum* and *Cytisus*, and this must be regarded as a chemical characteristic.

These crude origins, however, can hardly be considered part of the modern science of chemotaxonomy, and they have their limitations. Thus, while one might correctly predict that the similar fruits of the pea, runner bean and broad bean (genera *Pisum*, *Phaseolus* and *Vicia* respectively) were equally edible, it would be a fatal mistake to carry that conclusion as far as *Laburnum* and *Lupinus*, although all five genera are in the same family. Confusion between the swollen roots of the umbellifers *Pastinaca sativa* (parsnip) and *Oenanthe crocata* (water dropwort) has also caused a number of deaths, as has confusion between the leaves of the latter and those of another umbellifer, *Apium graveolens* (celery).

There are probably three main reasons for the recent rapid growth of chemotaxonomy: the development of many new techniques (notably various forms of chromatography and electrophoresis) which have made the analysis of plant products so much quicker and simpler, and which require much less plant material than formerly; the realization that behind the universal occurrence in plants of many vital biochemical pathways there is an enormous variation between taxa in many other, less vital pathways; and the current belief that evidence from as many sources as possible should be used in plant classification.

There now exists a vast bibliography on chemotaxonomy, most of it with reference to plants rather than to animals, but fortunately there are many authoritative reviews and compilations. Two encyclopaedic, multi-volume works, by Hegnauer[191] and Gibbs,[153] stand out in significance, while several textbooks and symposium volumes require mention.[6,25,37,180,187,233,257,431,432] There are journals (e.g. *Phytochemistry*, Vol. 1 in 1962) and serial reviews (e.g. *Advances in Phytochemistry*, Vol. 1 in 1968) devoted to the subject, the relatively elementary textbook by Smith,[385] and many excellent review articles.[129,182]

The identification of compounds isolated from plants is a practical problem which need not concern us here.[183] In the case of secondary metabolites it usually involves the separation of the crude mixture by some form of chromatography and the probable identification of the components by comparison with the patterns of chromatographic behaviour of known substances. This is what Smith[385] called the 'short-cut approach'; the alternative is the full chemical characterization of each component, which is preferable but out of the question in routine work. Some chemotaxonomists have sought to cut even more corners by not attempting any identification at all, but by comparing taxa on the presence or absence of various 'spots' in common on the chromatogram. This approach has yielded useful taxonomic data,[470] but its application is of limited value and could easily lead to false conclusions where the spots did not actually represent single, distinctive compounds.

The extent to which chemical data are used in classification varies enormously. Part of the reason for this lies in the expertise, facilities and man-hours which are necessary (despite statements to the contrary by some phytochemists) in order to carry out adequate investigations on a range of plant material. Without such full investigation the use of chemical information is very risky. There is also the problem that there are very few workers who are really highly skilled in both chemistry and plant taxonomy; almost all work is carried out by people trained in one field *or* the other, with obvious disadvantages.

Once these problems are overcome the contribution of phytochemistry to plant classification is enormous. There is a vast range of data which can be used to improve existing classifications and point the way to new ones. There is, however, no reason to suppose that chemical data are more important than structural data; the presence or absence of a particular chemical compound is not *a priori* a more (or less) valuable piece of taxonomic evidence than, say, the presence or absence of petals. Both are manifestations of the genotype of the organisms, and both equally prone to the short-comings and pitfalls associated with such characters. Sometimes it is the one, sometimes the other, which turns out to be the more indicative of taxonomic relationship. Attempts to utilize the genotype itself (DNA), or its immediate products (RNA and proteins), are discussed later in this chapter.

At present there are many groups of plants in which phytochemical data have contributed to substantial taxonomic improvements (some examples are given later), although this is not true for the great majority. Even where these data have been successfully utilized such information is very rarely incorporated into keys or standard descriptions in Floras, because it is not useful in identification without special facilities, etc. Sometimes, however, chemical characters can be observed by swift and simple operations, and the use in such cases of 'spot-tests' is a valuable routine procedure. How better to distinguish a carrot from a parsnip than by colour, smell and taste, or the lichens *Cladonia rangiferina* and *C. sylvatica* by their colour-reaction to KOH solution? In many cases modern chemotaxonomy has in fact served to define precisely and to strengthen the long-practised use of chemical characters rather than to uncover new ones. The use of scent to separate taxa in genera such as *Mentha*, *Stachys*, *Allium*, *Thlaspi* and many others is well known. In the sense that the results of chemotaxonomic studies are to be regarded at two separate levels (their widespread and essential use in the production of classifications, in the construction of phylogenies and in the identification of fragments or unrepresentative samples of plant material; and much less frequently where they can be used in the routine identification of plants) they are comparable with the results of microscopic anatomical studies.

Compounds useful in plant taxonomy

Although in theory all the chemical constituents of a plant are potentially valuable to a taxonomist, in practice some sorts of molecules are far more valuable than others. Apart from inorganic compounds, which are of relatively little use, three very broad categories of compounds can be recognized: *primary metabolites*, *secondary metabolites*, and *semantides*.

Primary metabolites are parts of vital metabolic pathways, and most of them are of universal occurrence, or at least occur in a very wide range of plants. Aconitic acid (first isolated from *Aconitum*) or citric acid (from *Citrus*), for example, participate in the Krebs (tricarboxylic acid) cycle, and are present in all aerobic organisms; the presence or absence of such compounds is therefore not of much systematic value. The same is true of the 22 or so amino-acids which are known to be constituents of plant proteins, or any of the sugars which figure in the photosynthetic carbon cycle, and so on.

In some cases, however, the quantities of such metabolites vary considerably between taxa, and this in itself can be taxonomically useful. For example, taxa in which universally occurring substances were first detected (such as the two above) often possess particularly large quantities of the molecules concerned, well above the amounts which participate in the essential metabolic pathways, often as food-storage materials. Sometimes such compounds are stored in a different form from that in which they are metabolized, for example sedoheptulose, a sugar constituting the carbohydrate food reserve of the genus *Sedum*, which as sedoheptulose diphosphate is part of the photosynthetic carbon cycle.

Secondary metabolites (or secondary plant products) perform non-vital (or at least non-universally vital) functions, and are therefore less widespread in plants. It is of course this restricted occurrence among plants which renders them valuable as taxonomic information. The most well-known groups of compound which have been utilized in this way include alkaloids, phenolics, glucosinolates, amino-acids, terpenoids, oils and waxes, and carbohydrates. There is much argument in the literature concerning the importance to the plant, i.e. the function, of secondary metabolites. Much of this discussion is, however, sterile because, while it is possible to demonstrate a function, it is impossible to prove that a particular substance has no function. And, as mentioned elsewhere, the emphasis placed on a character for taxonomic purposes should not be weighted by considerations of its function. Hence the extreme views that secondary metabolites all have some function or other (in many cases not so far discovered by man), or that they have no primary function but that many have acquired a function accidentally, or any compromise of these extremes, have little relevance to taxonomists, although they may be vital in physiological or evolutionary considerations. The value of such substances in taxonomy is judged simply by their distribution and correlation with other characters.

Secondary plant products are largely waste substances, foodstores, pigments, poisons, scents, structural units or water repellents, etc. In many cases they obviously do have an essential function, but of a general nature so that the precise molecular configuration of the compound is not vital. Thus a yellow pigment with an absorption maximum at 477 nm (which presumably defines its function) might be a betalain or an anthocyanin, and a poison might be an alkaloid or a glycoside.

Semantides are the information-carrying molecules. DNA is a *primary semantide*, RNA a *secondary semantide*, and proteins are *tertiary semantides*, following from the sequential transfer of the genetic code from the primary genetic information (DNA). In theory, the sequences of nucleotides and amino-acids in these substances should provide all the taxonomic information necessary for classification, and offer an alternative to the study of secondary metabolites, cytology, morphology and anatomy, etc., for the latter are merely manifestations of the former. However, in practice there are great difficulties in the gathering of the sequence data, and the results frequently present puzzling anomalies. Thus different classes of proteins may show different levels of similarity from the same pairs of organisms, and there can be no hope that semantide sequence studies will provide the ultimate taxonomic criterion in the foreseeable future, even if means were found of applying the techniques to all taxa of organisms. Sometimes the semantides, together with the larger polysaccharides, are known as *macromolecules*, and the primary and secondary metabolites as *micromolecules*.

Most chemical substances which have been found very useful in taxonomy are therefore secondary metabolites or semantides. In the former case they are usually rather large molecules with many side-groups which can be variously substituted, thus allowing a wide range of possible types of molecule, a good proportion of which is likely to be of limited occurrence. Many of them are, for this reason, the end-products of metabolic pathways, or parts of short side-chains from widely distributed metabolic pathways. The greater the complexity of a molecule the greater the number of steps required for its formation and therefore the narrower its distribution—hence the greater its taxonomic value. Semantides, on the other hand, provide taxonomic data not on the basis of presence or absence, but in terms of sequences, ratios or percentages.

The problem of convergence in chemical compounds is as great as that in structural characters. Many common chemical substances in plants are parts of different metabolic pathways, and their presence in two taxa might be due to the presence of quite different sets of enzymes. Obviously, the more complex a molecule and the further removed it is from a common metabolic pathway the less likely it is to be polyphyletic, but in many cases it is simply not known what metabolic patterns lie behind its presence. The possession of various sulphur-containing compounds of great diversity, with their characteristic smells, is a well-known feature of the monocotyledon genus *Allium* (onions, etc). Certain other genera, for example *Milula*, possess a similar smell and other characteristics in common, and the presence of the chemical substances involved surely strengthens the likelihood that the two genera are closely related (they are placed in the same family in some classifications but in different ones in others). On the other hand the same smell, and similar substances, also occur in various Fabaceae and Brassicaceae, for example *Alliaria* and *Thlaspi alliaceum*, which are dicotyledons obviously far removed from *Allium*. Clearly, in the latter case the chemical characters do *not* indicate a close relationship. These situations are clear-cut, but it would be very difficult to assess the evidence if the same chemical compounds were found in genera of monocotyledon families (such as Iridaceae and Dioscoreaceae), which are obviously quite distinct from *Allium* but which might conceivably be more closely related than hitherto considered. There are many unsolved problems of this nature in chemotaxonomy, and many compounds which are already known to be polyphyletic. The evidence for several separate origins of tropane alkaloids has been presented by Romeike.[364]

Problems such as these lead one to the conclusion that the best characters of secondary metabolites involve the presence of chemical *pathways* rather than chemical *substances*, the former indicating the possession of a *series* of enzymes rather than a single compound.[35] If one could unravel the metabolic pathways behind every type of molecule one could use single compounds (which were monophyletic) as indicators of whole pathways. However, it is not likely that we shall be in a postition to make such prognostications for a very long time; moreover, the taxonomically haphazard occurrence of metabolic sequences such as the C4 photosynthetic pathway and Crassulacean Acid Metabolism[69] suggests that even the use of pathways rather than compounds will not provide all the answers.

Value of chemotaxonomy

Chemotaxonomic characters, like any others, are useful at all levels of the taxonomic hierarchy, and it is not possible to predict the level of their value. Simple examples can be drawn from colour characters. Flower colour often varies greatly within one species; especially where garden varieties have been selected, red, blue, yellow, white and all combinations and intermediates can exist in one species. However, in some genera flower colour is of great importance at the specific level. *Tragopogon porrifolius* has purple flowers, whereas *T. pratensis* has yellow (hybrids have a patchy distribution of the colours); *Silene alba* has white petals whereas *S. dioica* has red (hybrids have pink petals); *Medicago sativa* has purple petals, whereas *M. falcata* has yellow (hybrids have either a patchy distribution of colours or an intermediate colour, i.e. green); *Hyacinthoides non-scriptus* has cream anthers, whereas *H. hispanicus* has blue (hybrids have pale blue anthers). In all these four pairs the two species are closely related (they all hybridize), and these colour differences are important because there are few discriminating characters. In some of these examples there is colour variation, which complicates the issue. Thus albino variants of *Silene dioica* are much less easy to distinguish from *S. alba*, and albino variants of the two species of *Hyacinthoides* (bluebell) are likewise more similar than blue-flowered plants. In the latter genus pink-flowered variants are distinguishable by colour, since the anthers are also pink in *H. hispanicus* but cream in *H. non-scriptus*. In some families, for example Apiaceae, flower colour can be an important generic characteristic. In *Adonis* (Ranunculaceae) some species are always red-flowered, some always yellow-flowered, and some (e.g. *A. flammea*) can be either colour. Anther colour, so important in the genus *Hyacinthoides*, is of very little taxonomic value in the Poaceae, where a very wide range of species in different tribes, including many of the common European grasses, have variants with yellow and with purple anthers. This seems likely to be a simple genetic variation which exists throughout a whole family.

A final example concerning colour can be used to illustrate the use of chemical characters at the upper levels of the hierarchy. The primary and accessory photosynthetic pigments (chlorophylls, carotenoids and biliproteins) have extremely distinctive distributions in photosynthetic organisms. In all of them the primary pigment is chlorophyll *a*, except in bacteria, where it is one of the bacteriochlorophylls (Table 4.1). This difference is associated with a difference in the photosynthetic mechanism, for organisms with chlorophyll *a* all split water and release oxygen ($H_2O \rightarrow 2H^+ + 2e^- + \frac{1}{2}O_2$), whereas this is never the case with bacteria. In the latter the photosynthetic substrate is varied; it may be H_2S, with the evolution of sulphur instead of oxygen, or organic. These characteristics set the Cyanophyta (blue-green algae) apart from bacteria, although most modern classifications place them together in the Procaryota (as Cyanobacteria) because of their distinctive ultrastructure. These data suggest that, in terms of the endosymbiotic hypothesis (p. 187), the Procaryota that gave rise to plant chloroplasts were allied to blue-green algae rather than to bacteria.

The other chlorophylls and the other pigment groups differ characteristically between the eleven divisions of algae named in the Appendix. Chlorophyta and Euglenophyta share a very similar set of pigments, and the same set is also found in all higher plants (Embryobionta). This is one of the main lines of evidence suggesting that higher plants and green algae must have a common ancestry, different from that of other algal groups. The disposition of these pigments among the algae (Table 4.1) is nowdays one of the major characters used in algal taxonomy. It has greatly helped in decisions on the assignations of various algae to their correct division, for example *Vaucheria* to the Xanthophyta (not to the Chlorophyta as previously). The occurrence of biliproteins is far more limited than that of the other two pigment classes, for they are restricted to the blue-green algae, Cryptophyta and Rhodophyta. This is a further character linking the blue-green algae with eucaryotic algae rather than with bacteria.

Chemical examples of discrimination between algal divisions and classes could have been almost equally well taken from food-storage compounds or cell-wall components (Table 4.1). In fact chemical characters have been very prominent in algal taxonomy for a long time, well before the present upsurge in the chemotaxonomic study of vascular plants, and this illustrates the very varied extents to which this sort of evidence has been utilized in different taxonomic groups. At one extreme are the bryophytes, in which very few chemical analyses have been made on a systematic basis and on which chemotaxonomy has therefore made very little impact.[215,430] At the other extreme are the lichens, in which chemical characters have been utilized as standard taxonomic procedure for over a century, mostly by the use of simple spot-tests. In the lichens, taxa are sometimes separated by such tests alone, but more sophisticated phytochemical studies are now being increasingly pursued.[81,82,83,376] In the bacteria, to use another example from outside the plant kingdom, *most* of the taxonomic criteria are chemically based, a necessity in a group where distinctive structural features are so few.

Table 4.1 The major pigments, storage carbohydrates and cell-wall compounds of procaryotes and plants. In the pigment column the presence of chlorophylls (chlor), carotenes, xanthophylls and biliproteins is given in that order; in the case of xanthophylls the most important ones are named and an indication of the numbers of others is provided.

Taxon	Major pigments	Major storage carbohydrates	Major cell-wall compounds
Bacteria	Bacteriochlorophylls a, b, c, d; various carotenes and xanthophylls (not α- or β-carotene)	Starch, glycogen, poly-β-hydroxybutyric acid	Mucopeptides, polysaccharides, lipopolysaccharides
Blue-green algae	Chlor a; β-carotene; myxoxanthophyll + 3; biliproteins	Cyanophycean starch	Mucopeptides, polysaccharides, lipopolysaccharides
Rhodophyta	Chlor $a + d$; α- + β-carotene; lutein + 2; biliproteins	Floridean starch	Cellulose, hemicelluloses
Cryptophyta	Chlor $a + c$; α-carotene; alloxanthin; biliproteins	Starch	Cellulose, hemicelluloses
Pyrrophyta	Chlor $a + c$; β-carotene; peridinin + fucoxanthin + 1	Starch	Cellulose, hemicelluloses
Xanthophyta	Chlor $a + c$; β-carotene; diadinoxanthin	Leucosin	Cellulose, hemicelluloses
Chrysophyta	Chlor $a + c$; α- + β-carotene; diadinoxanthin + fucoxanthin + 1	Leucosin	Hemicelluloses, silica
Bacillariophyta	Chlor $a + c$; β-carotene; diadinoxanthin + fucoxanthin + 1	Leucosin	Hemicelluloses, silica
Phaeophyta	Chlor $a + c$; β-carotene; fucoxanthin + 1	Laminarin	Cellulose, hemicelluloses
Chlorophyta	Chlor $a + b$; α- + β-carotene; lutein + neoxanthin + 1	Starch	Cellulose, hemicelluloses
Prasinophyta	Chlor $a + b$; α- + β-carotene; neoxanthin + 1	Starch	Cellulose, hemicelluloses
Charophyta	Chlor $a + b$; β-carotene; lutein + neoxanthin + 2	Starch	Cellulose, hemicelluloses
Euglenophyta	Chlor $a + b$; β-carotene; neoxanthin + 1	Paramylon	None
Embryobionta	Chlor $a + b$; α- + β-carotene; lutein + many	Starch	Cellulose, hemicelluloses, lignin

Recently, some strange bacteria living in various extreme environments (e.g. anaerobic conditions, high salinity, high temperature) have been found to differ from all other Procaryota in a range of chemical characters (e.g. lipids based on glycerol ethers rather than on glycerol esters, distinctive RNA base sequences, unique intermediary metabolism often poor in energy-production, different cell-wall constituents). Despite their diverse morphology and habitats they have been given taxonomic recognition as Archaebacteria. This taxon is thought by many to be the most primitive one in the living world, and one to be maintained at a higher rank even than the Procaryota/Eucaryota distinction. Its recognition is based *solely* on chemical data.

In gathering chemical data full account must be taken of non-taxonomic variation. Variation in the quantity of a substance present can lead to totally wrong conclusions. Probably many of the earlier reports of the absence of a compound in fact mean only that it was not detected with the methods available. The techniques of today are far more sensitive, but one cannot be certain that they are always able to detect minute quantities of the substance being sought. The enormous range in the alkaloid content of various strains of *Papaver somniferum* (opium poppy) is well known, but there are many other examples as well; surveys based on single or few samples of a taxon might not be very reliable. There can also be variation according to the parts of a plant sampled[126] (e.g. the edible and poisonous parts of potatoes and tomatoes), the conditions under which it has been growing (e.g. the 'greening' of exposed potato tubers associated with the development of solanine), the stage of the life cycle or the season in which it is sampled (e.g. the seasonal differences in isozyme patterns in *Juniperus*[247]) and the conditions under which the plant material has been stored or its chemical constituents extracted (e.g. the poisonous component of *Ranunculus* is lost on drying, whereas that of *Helleborus*, also Ranunculaceae, is not). These examples serve to indicate that some degree of standardization is needed in the gathering and presentation of chemical data for taxonomic purposes. On the other hand a comparison of carbohydrates found in wild-collected hepatics with those found in the same species grown in sterile culture showed no qualitative differences.[62]

Examples from secondary metabolites

Of all the groups of secondary metabolites used by chemotaxonomists probably none has provided more taxonomic data than *phenolic compounds*. These form a very loose class of compounds having in common only the fact that they are based upon phenol, C_6H_5OH (Fig. 4.1A). Most of them are far more complicated in structure than phenol itself, having several aromatic rings and several substitution groups or side-chains. Many of them have no known functions in plants, but some are the most important flower pigments and others are involved in the inhibition of pathogenic fungi. General reviews have been written by Bate-Smith,[18] Harborne[178,179] and Ribereau-Gayon.[355]

The taxonomically most important phenolics are the *flavonoids*, which have a relatively common nucleus (Fig. 4.1B) with a great variety of types and patterns of side-groups which characterize the individual compounds. There is usually a considerable diversity of flavonoids in any one species; some of these are very widespread, others very rare, and the pattern and combination

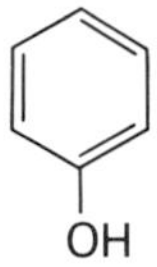

A Phenol

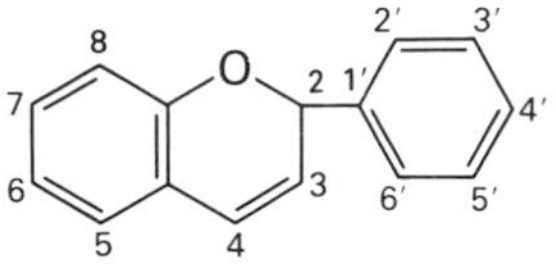

B The flavonoid nucleus

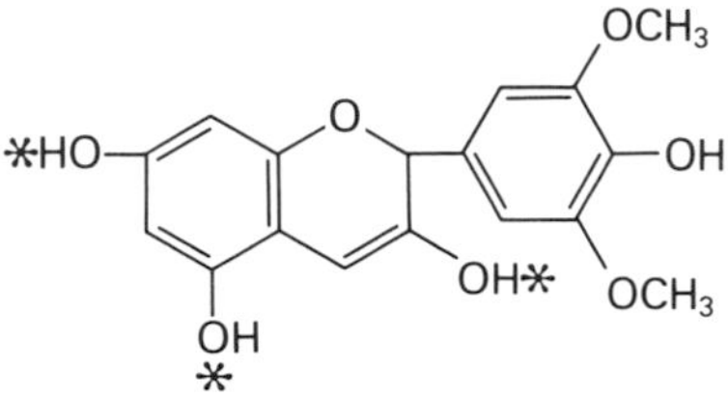

C Malvidin

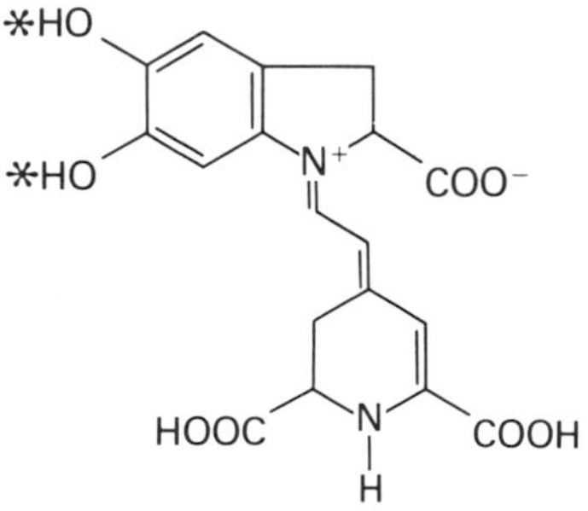

D Betanidin

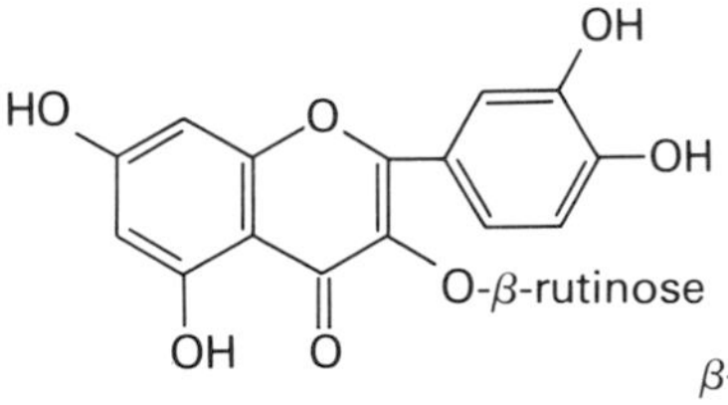

E Quercetin-3-β-rutinoside
from *Baptisia leucantha*

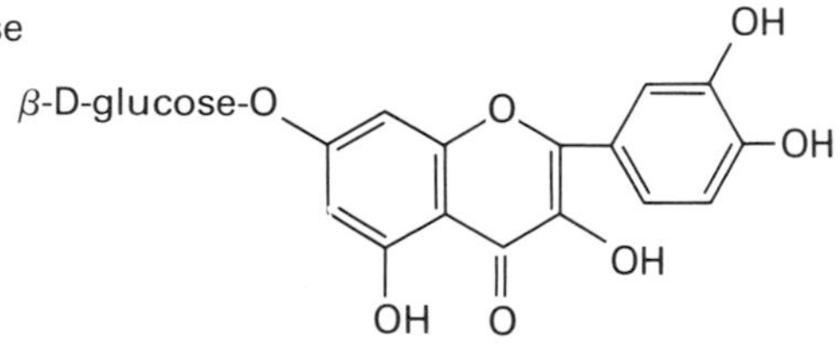

F Quercetin-7-β-D-glucinoside
from *Baptisia sphaerocarpa*

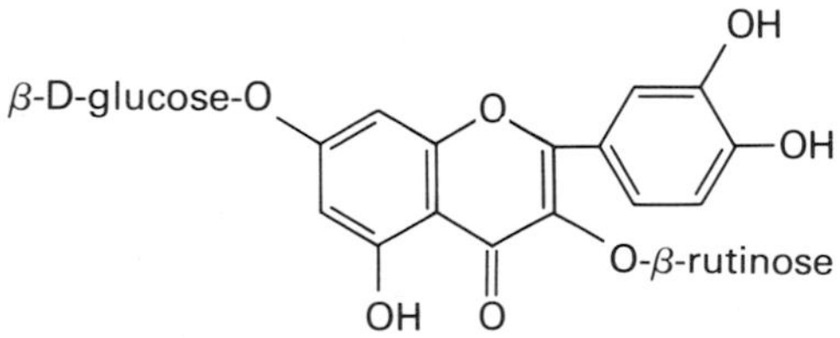

G Quercetin-3-β-rutinosyl-7-β-D-glucinoside from
Baptisia leucantha × *B. sphaerocarpa* (absent
from both parent species)

Fig. 4.1   Structure of various phenolic compounds mentioned in the text.

of occurrences have on many occasions proved valuable as taxonomic evidence in the flowering plants at all levels from order downwards.

One of the best examples of the taxonomic value of secondary metabolites concerns flower pigments. Most red, blue and similar colours of flowers and other organs indicate the presence of **anthocyanidins**, a particular class of flavonoid. Malvidin (Fig. 4.1C) is a good example. The hydroxyl groups at positions 3, 5 and/or 7 (asterisked in Fig. 4.1C) are frequently substituted by a sugar (such as glucose or rhamnose), the resultant compound being known as an anthocyanin. The combination of the types of anthocyanidin and types and position of sugars attached to them provides a large number of anthocyanins, which are extremely widely distributed in flowering plants (in almost all families). They are absent, however, from a few families of dicotyledons, where their function is taken over by a quite unrelated class of compounds, the **betacyanins**. These compounds differ conspicuously from anthocyanidins in that heterocyclic nitrogen-containing aromatic rings are present, and their synthesis in the plant is along quite different metabolic pathways. Betanidin (Fig. 4.1D), from beet (*Beta vulgaris*—hence the chemical name), is an example. They resemble anthocyanidins, however, in that they appear to carry out the same functions and they may also be attached to sugar molecules at one of two positions (asterisked in Fig. 4.1D). Closely related to betacyanins are **betaxanthins**, which are yellow pigments apparently with similar functions to various yellow or cream flavonoids, loosely known as **anthoxanthins**, which are present in most plants.

The important taxonomic evidence from these chemical data stems from the *mutually exclusive* nature of the betacyanins and betaxanthins (collectively **betalains**) on the one hand and the anthocyanins on the other, although other flavonoids (some of them anthoxanthins) may co-exist with the betalains. It must be emphasized that the presence of anthocyanins or betalains is not a single character, but represents the presence of one metabolic pathway and the absence of another, together involving, presumably, a difference in a good many enzymes (and genes). Moreover it is not easy to envisage a frequent interconversion between the two pigment systems, and for this reason the presence or absence of betalains has been taken as a very reliable taxonomic character. It should be mentioned, however, that betalains are now known also from certain basidiomycete fungi,[268] in some cases the same substance occurring in these and in the flowering plants. Betalains have been found in nine families of dicotyledons, all members of, or related to, the old order Centrospermae, which was delimited on various morphological and anatomical characters, particularly of the embryo. Thus the newer definition of the Centrospermae differed from the old (Table 4.2) in the exclusion of the Caryophyllaceae and Molluginaceae, which possess anthocyanins and lack betalains, and in the inclusion of the Cactaceae, which had usually been placed in a separate order but which were found to contain betalains. Although the structural data do not argue strongly against the inclusion of the Cactaceae, the exclusion of the Caryophyllaceae and Molluginaceae is in direct contradiction to the anatomical evidence. Thus many taxonomists argued for one system or the other according to which evidence they favoured. These arguments have recently been complicated by the discovery (mentioned in Chapter 3) of the ultrastructural characters of the plastids of

Table 4.2 Classifications of the order Centrospermae (Caryophyllales) varying according to the use of evidence from pigments.

Structural classification	Chemical classification	Compromise classification
Centrospermae	*Chenopodiales*	*Caryophyllales*
Aizoaceae	Aizoaceae	*Chenopodiineae*
Amaranthaceae	Amaranthaceae	Aizoaceae
Basellaceae	Basellaceae	Amaranthaceae
Caryophyllaceae	Cactaceae	Basellaceae
Chenopodiaceae	Chenopodiaceae	Cactaceae
Didiereaceae	Didiereaceae	Chenopodiaceae
Molluginaceae	Nyctaginaceae	Didiereaceae
Nyctaginaceae	Phytolaccaceae	Nyctaginaceae
Phytolaccaceae	Portulacaceae	Phytolaccaceae
Portulacaceae		Portulacaceae
Cactales	*Caryophyllales*	*Caryophyllineae*
Cactaceae	Caryophyllaceae	Caryophyllaceae
	Molluginaceae	Molluginaceae

the sieve-tubes (Fig. 3.5), for a particular type of plastid (with peripheral ring-shaped bundles of proteinaceous filaments, termed type P III (Fig. 3.5D)) precisely characterizes the betalain-containing families *together with* the Caryophyllaceae and Molluginaceae. This has been taken to show that, whereas the nine betalain-containing families should be kept taxonomically distinct from the other two, they should be placed close to them, either as the only two orders in a superorder or subclass, or the only two suborders in an order (Table 4.2). On the other hand, some taxonomists are still maintaining that the anthocyanin- and betacyanin-containing Centrospermae do not represent distinct taxa at any level, but that the two families of the former group have closest affinities with different betacyanin-containing families (but see p. 108).

It is rare that chemical data, even with support from other sources, supply such significant taxonomic evidence as the above example, but a summary of this rather well-worn story[152,269] serves to show the potential value of chemotaxonomy. Many other good examples of its value at the level of family and above in the angiosperms are available.[152] Illustrations of its use at infraspecific levels and in hybridization studies may be taken from work with terpenoids as well as with phenolics. The great variation shown by some species of *Eucalyptus* in terms of essential oil content, leading to the recognition of morphologically identical chemical races, is well-known, as is the presence of cyanogenic and non-cyanogenic races of *Trifolium repens* and *Lotus corniculatus*. In the genus *Xanthium* (Asteraceae) the different ecotypes of one species have been shown to vary in their sesquiterpene lactones.[277] Cryptic chemical races of *Conocephalum conicum* with different flavonoids occur in Asia, Europe and America,[328] and many more examples of this kind of variation are coming to light.

In other instances the variation is not disjunct, but continuous (i.e. clinal, see Chapter 7). For example, Flake *et al.*[135] analysed samples of *Juniperus virginiana* every 150 miles along a 1 500 mile north-east/south-west transect from Washington D.C. to Texas, USA, for the presence of a wide range of terpenoid compounds, and detected a gradual change in terpenoid composition along the transect. Adams and Turner[4] made similar studies on the related *J. ashei* throughout its range of distribution in Texas, and found that the more peripheral populations showed greater divergence from the population centre. These studies are particularly interesting because *J. ashei* and *J. virginiana* had been the subject of much detailed morphological and statistical work,[174] which had strongly indicated that the cause of at least some of the variation in *J. virginiana* was its hybridization with *J. ashei* at and towards their point of contact in Texas. The clinal variation of *J. virginiana* (which is also apparent in morphological features) was attributed to the greater and greater dilution of the effects of hybridization (i.e. the lower and lower frequency of *J. ashei* genes) away from the south-western limits of *J. virginiana*. This dilution effect, which has many well substantiated examples, is known as ***introgression*** (see Chapter 6). Although specifically searched for, no chemical evidence of introgression or hybridization between the two junipers was encountered by the two groups of workers mentioned above.

Nevertheless, chemotaxonomy is a discipline very well suited to the detection of hybridization, and there are many examples where it has been very successful. The best documented research in this field is that by B. L. Turner and his associates in the legume genus *Baptisia*. In general, chemical compounds are inherited *additively* in hybrids, i.e. an F_1 hybrid tends to possess the enzymes and hence the molecular products of both its parents. Just as the offspring of crosses between plants with white and coloured petals usually possess coloured petals (albeit often of a paler shade), so also do the invisible compounds behave; such a pattern of inheritance is totally as would be expected. Studies on the flavonoids of *Baptisia* have shown that each species is well characterized by a distinctive spectrum of flavonoids (Fig. 4.2), and that hybrids can be easily detected in mixed populations. In one field in Texas, Alston and Turner[7] discovered four species of *Baptisia* as well as examples of all six possible binary hybrid combinations. It is most instructive to note that Alston and Turner found that it was impossible to differentiate between all these ten taxa on morphological characters alone, or on biochemical characters alone, but that a combination of the two disciplines effected

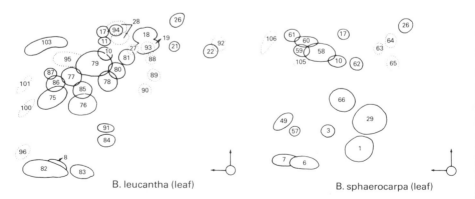

Fig. 4.2 Diagrammatic representation of chromatograms of the leaf phenolics of *Baptisia leucantha* and *B. sphaerocarpa*, redrawn from Alston and Turner.[5]

complete separation. Of the 125 flavonoids used by Alston and Turner many are species-specific markers which enabled them to validate a hybrid origin of plants which were not exactly intermediate between their parents, but were the result of backcrossing or of introgression.

Similarly, the additive inheritance of chemical compounds has been useful in the investigation of the parentage of **amphidiploids** (polyploid species of hybrid origin but which now behave as diploids, see Chapter 6). The classic example of this is the study by Smith and Levin[383] on the flavonoids of the ferns known as the 'Appalachian *Asplenium* complex'. This complex consists of three diploids and three tetraploids, all behaving as diploid species, as well as a number of sterile hybrids. The ancestry of the tetraploids and sterile hybrids in terms of the three diploid species was worked out in full by flavonoid analysis; in this case (cf. *Juniperus* above) the findings completely corroborated the conclusions previously drawn from morphological and cytological work.

As well as possessing the compounds of both parents, hybrids sometimes anabolize new substances present in neither parent. This is again an expected phenomenon which might arise in several ways, for example the completion of missing steps in an enzymic sequence (a form of genetic complementation between deficient mutants), or the processing of the same molecular skeleton by two different enzyme systems. Alston *et al.*[5] found such new sorts of flavonoids in hybrids of *Baptisia* species, for example a new quercetin glycoside in hybrids between the white-flowered *B. leucantha* and the yellow-flowered *B. sphaerocarpa* (Fig. 4.1E, F, G).

Because chemical compounds are such *simple* characters (the presence or absence of a precisely definable substance), they are useful in phylogenetic (as opposed to taxonomic) studies. Within any given anabolic pathway the presence of more complex substances must in general indicate greater evolutionary advancement (although their absence could be due to loss rather than to primitive non-attainment). Such reasoning has been used to trace phylogenetic lineages, ecological adaptations, geographical migration and differentiation and other rapidly evolving infraspecific variation in a very wide range of plants. Many detailed examples are given by Harborne[180] and in the elegant work on the turpentine components of *Pinus* by Mirov,[294] who was able to conclude that the genus originated in high northern latitudes and to follow its migration routes southwards.

Examples from semantides

Semantides are popular sources of taxonomic information because they comprise either the primary genetic information itself (DNA), or the secondary (RNA) or tertiary (protein) derivatives of it. It can be argued that the closer to the actual genetic material one investigates, the more fundamental are the data obtained and the more straightforward and uncomplicated should be their taxonomic interpretation; also that the failure to date of such data to provide the bulk of at least chemotaxonomic evidence (if not of all systematic evidence) is simply due to practical difficulties. In fact nucleic acids have provided a negligible amount of taxonomic information in plants so far, but proteins are probably second only to phenolic compounds in the degree to which they have been utilized by chemotaxonomists. There are several comprehensive reviews.[37,43,127,128,129,161,233]

The results obtained from protein taxonomy are largely divisible into three main headings, corresponding to the three main methods employed: *serology*; *electrophoresis*; and *amino-acid sequencing*.

The technique of serology relies on the immunological reactions shown by mammals when they are invaded by foreign proteins, phenomena first demonstrated in 1897 and utilized by plant taxonomists soon after. Put very simply, if a plant (or animal or microbial) extract (A) containing proteins (*antigens*) is injected into a mammal (usually a rabbit), the mammal will form proteinaceous *antibodies*, each specific to an antigen and capable of coagulating it (and hence rendering it non-functional). These antibodies can be extracted from the blood of the animal as an *antiserum*. Since the latter will coagulate further supplies of antigens it can be used as a standard test against other plant extracts (B, C, D, etc.) and the amount of coagulation which it causes in them can be used as a measure of their similarity to plant extract A (and hence the similarity of species B, C, D, etc. to species A). Early work was very crude but became more reliable by the use of more refined sampling and recording techniques.[96]

The individual antigen–antibody reactions, together forming the total precipitation, can be recorded separately, either by allowing the antigenic material and antiserum to diffuse towards one another in a gel (the different proteins travelling at different rates, and hence the different reactions occurring at different positions in the gel), or by first separating the antigens unidimensionally in a gel by electrophoresis, and then allowing these to diffuse towards a trough of antiserum (Fig. 4.3). One advantage of the former method (*double-diffusion serology*) over the latter (*immuno-electrophoresis*) is that the precipitation reactions of several different antigen mixtures (from different taxa) to a given antiserum can be observed at the same time on a single gel, but the separation of the constituent reactions is often better with the latter method. Nowadays there are many modifications of these two major techniques; for example, in double-diffusion serology the antiserum is more usually placed in a small circular well surrounded by a ring of similar wells containing the samples of antigens. A refinement known as *absorption* involves the prior removal of common antibodies in the antiserum, so that the remaining ones can be compared more easily with those of other taxa. Other refinements include radio-immunoassay (RIA) and enzyme-linked immunosorbent assay (ELISA), where either the antibodies or the antigens are labelled with radioactive molecules or enzymes, enabling their detection when in very small quantities.

Serology does not usually involve the *identification* of particular proteins; often one is comparing unknown proteins whose general function (food-storage, structural, enzymic, etc.) might be unknown. This can be a disadvantage in interpreting the significance of the results, and attempts are sometimes made at precise identification by running parallel electrophoretic separation on similar samples. Serological investigation of a range of tissues has been made, for example leaves, fruits, seeds, tubers, spores and pollen. Different parts of a plant would be expected to yield different proteins, as different

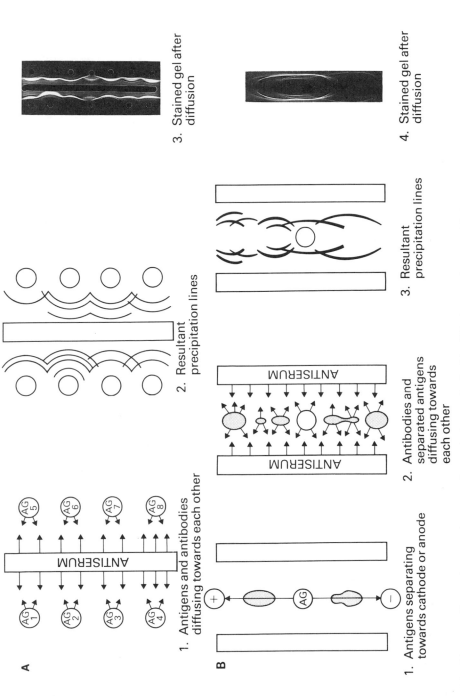

Fig. 4.3 Diagrams illustrating protein-separation techniques: **A,** double-diffusion serology; and **B,** immuno-electrophoresis. The photographs, taken from Smith,[384] show actual results with various species of *Bromus* using antiserum to *B. hordeaceus*.

enzymes are present, and different proteins (or none at all) may be used as food-reserves, although some proteins of course occur in all parts of the plant. Most work has been concentrated on organs where protein is found as a food-reserve, such as seeds[453] and stem tubers.[188]

Serology has proved a useful taxonomic tool at all levels from above the family to below the species. Some of the best-known work has been carried out by Hawkes and his students[188] on the relationships of the tuber-bearing species of *Solanum* (potatoes), aimed partly at elucidating the ancestry and close relatives of the cultivated potato. At and around the familial level the work of Jensen[231,232] on the Ranunculaceae and Berberidaceae, and that of Fairbrothers and co-workers on families related to the Cornaceae[127] and Magnoliaceae,[130] have yielded many valuable data. In the former case the classification of the Ranunculaceae at tribal and generic levels indicated by serological relationships strongly supported that based on classical evidence, particularly from the chromosomes. At lower levels Smith[384,387] has studied the relationships of annual brome-grasses (*Bromus*), relating immuno-electrophoretic data to those from morphology, cytology and cytogenetics. He was able to throw new light on species problems, diploid–tetraploid relationships and hybrids, and the serological and cytological distinctness of a previously named variety of *Bromus secalinus* led him to recognize it as a new species, *B. pseudosecalinus*. Prus-Głowacki[333] was able to demonstrate the extent of introgression between *Pinus sylvestris* and *P. mugo* using immuno-electrophoretic techniques.

Electrophoretic separation of proteins relies on their amphoteric properties, whereby they are charged positively or negatively to various extents according to the pH of the medium, and will travel through a gel at various speeds across a voltage gradient. This is usually carried out in a column of acrylamide gel. The column is composed of two discontinuous gels (hence the term *disc electrophoresis*) placed contiguously, a gel of larger pore size being situated above a gel of smaller pore size. Protein separation depends upon the sieving effect of the gel pores, a crude separation being effected in the larger pored zone and a complete separation into discrete bands in the smaller pored zone. After a suitable separation time, the current is switched off and the proteins located by dyes; general protein stains may be used or, if a particular enzyme is being studied, a specific stain. An alternative technique is *isoelectric focusing*, where a gel of single pore size is set up with a pH gradient (say of 3–10) and, when the extract is applied and the current switched on, the proteins come to lie at the position in the gradient corresponding with their isoelectric point. This has the advantage that subsequently the proteins in the gel can be separated on a plate in a second dimension by standard disc electrophoresis, relying on molecular size rather than on charge. By use of this two-dimensional method over 1 000 proteins have been separated from one organism.

The taxonomic applications of the results of electrophoresis have been useful mainly at and below the generic level. While much work has been directed at food-reserve proteins (e.g. in legumes and cereals), more has involved enzymes. Not only can electrophoresis separate closely related storage proteins and enzymes, but it has also led to the recognition of *allozymes* and *isozymes*, which are different forms of what were previously considered single enzymes. Allozymes are different forms of an enzyme where the constituent polypeptides are determined by different alleles at one locus; isozymes (or isoenzymes) are different forms where the polypeptides are determined by more than one locus. The former appear to be the more common, but both are loosely termed isozymes in much of the literature. These isozymes or allozymes can be distinguished by their different electrophoretic mobilities and are thought to differ also in the precise conditions needed for their optimal catalytic activity; hence they might be brought into operation by the plant in different organs, at different stages of growth, or in different habitats. Enzyme polymorphism is often exhibited below the species level and, especially when the constitution of more than one enzyme is worked out, isozymes can be used to recognize even individual clones. They are therefore of use not only to taxonomists, but also to ecologists and population biologists. In some cases inter-populational variation in isozymes is accompanied by morphological differentiation, for example enzyme races of *Conocephalum conicum* in Poland.[435]

The most valuable electrophoretic results have perhaps been obtained in the cereals related to wheat, in relation to the genomic constitution and ancestry of the tetraploids and hexaploids. For example, Johnson,[234] working on storage proteins, concluded that the hexaploid bread-wheat (*Triticum aestivum*) did indeed contain a sum of the proteins possessed by the diploid species that had been postulated on morphological and cytological evidence to be ancestral to it (Fig. 4.4). Barber,[15] studying enzymic proteins, found that certain polyploids possessed the isozymes of all their progenitors, plus some new ones. These new hybrid isozymes are presumably oligomers incorporating polypeptides in new combinations of genomes in the polyploids. Much evidence is now available of the value of isozyme analysis in the study of breeding systems, population genetics, polyploidy, hybridization and phylogeny.[162,216,424,487]

Amino-acid sequencing is a more modern development, aiming to identify pure proteins down to the atomic level. It is possible to break off the amino-acids from the polypeptide chain one by one, identify each in turn chromatographically, and so build up the complete sequence of amino-acids step by step. A first step is to break the total polypeptide chain into smaller peptidic fragments, each of which is then sequenced separately. In the past these small peptidic units were separated by chromatography and/or electrophoresis (*fingerprinting*), and such relatively crude results themselves proved of taxonomic value. Today, complete sequencing is a highly automated routine procedure. Amino-acid sequencing investigates the variation in the precise sequence of amino-acids in a single homologous protein (i.e. one presumably of monophyletic origin) throughout a range of organisms. This relies upon the fact that a particular protein does not have a single invariable structure, but a good proportion of it may vary without altering its essential function (cf. isozymes). In the case of cytochrome *c*, for example (the molecule first used in plants for sequencing purposes), about 79 out of the approximately 113 amino-acids vary from species to species, but alteration of even one of the other 34 destroys the functioning of the molecule. Cytochrome *c* is an ideal molecule as it is relatively small and stable, is coloured, and is ubiquitous in aerobic organisms. So far its sequences in at

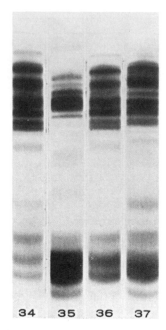

34   35   36   37

Fig. 4.4 Electrophoretic profiles of seed proteins of *Triticum dicoccum* (tetraploid 34), *Aegilops squarrosa* (diploid 35), a mixture of extracts from the last two (36), and *Triticum aestivum* (hexaploid 37), taken from Johnson.[234] *T. aestivum* is thought to have arisen by hybridization between the other two species, and its profile is closely mimicked by that of the mixture.

least 25 species of vascular plant have been determined (Fig. 4.5), besides others in algae, animals, fungi and bacteria.

Other widely-occurring proteins that have been sequenced for taxonomic or phylogenetic purposes include plastocyanin, ferredoxin and ribulose-1.5-bisphosphate carboxylase. The plastocyanin sequence of over 70 species of angiosperms has been determined. Probably the most valuable results will be obtained by using data from several different macromolecules rather than just one.[272] Another approach is to use data from one macromolecule obtained from a range of methods, e.g. the sequencing, isoelectric focusing, electrophoresis and serology of ribulose-1.5-bisphosphate carboxylase.[165,273,475]

In general, the number of amino-acid differences is roughly parallel to the distance apart of the organisms in traditional classifications, suggesting that the method is broadly reliable.[43,45] However, there are some anomalies, which indicate that this measure of protein structure is not an infallible guide to relationship. For example, the number of differences between the cytochrome *c* of maize and wheat (two grasses) is greater than that between maize and some dicotyledons. In practice, these problems are magnified because almost all the work on amino-acid sequences has been aimed at unravelling phylogenies rather than at producing classifications. It is often assumed that evolution of the molecule has taken place by a minimum number of mutations (principle of parsimony), i.e. that there is no convergent evolution or back-mutation and that different positions on the molecule are

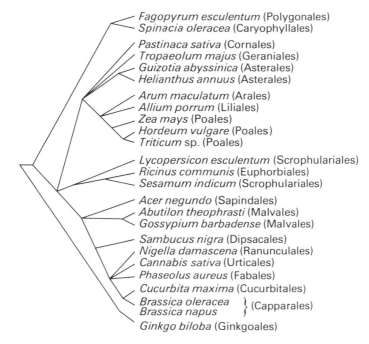

Fagopyrum esculentum (Polygonales)
Spinacia oleracea (Caryophyllales)
Pastinaca sativa (Cornales)
Tropaeolum majus (Geraniales)
Guizotia abyssinica (Asterales)
Helianthus annuus (Asterales)
Arum maculatum (Arales)
Allium porrum (Liliales)
Zea mays (Poales)
Hordeum vulgare (Poales)
Triticum sp. (Poales)
Lycopersicon esculentum (Scrophulariales)
Ricinus communis (Euphorbiales)
Sesamum indicum (Scrophulariales)
Acer negundo (Sapindales)
Abutilon theophrasti (Malvales)
Gossypium barbadense (Malvales)
Sambucus nigra (Dipsacales)
Nigella damascena (Ranunculales)
Cannabis sativa (Urticales)
Phaseolus aureus (Fabales)
Cucurbita maxima (Cucurbitales)
Brassica oleracea }
Brassica napus   } (Capparales)
Ginkgo biloba (Ginkgoales)

Fig. 4.5 Cladogram relating the cytochrome *c* amino-acid sequences of 25 species of spermatophyte, constructed using the 'ancestral sequence method', redrawn from Boulter.[43] Note the isolated position of the only gymnosperm, *Ginkgo biloba*.

equally liable to substitution. It is, of course, not possible to verify these assumptions (and, indeed, they seem unlikely to be fully justified). It is now known that plastocyanin and cytochrome *c* can exhibit a large number of parallel substitutions, i.e. identical changes from one amino-acid to another at the same position in the protein may occur independently in different organisms. In fact it is thought that in angiosperms both the latter two proteins have about 40% parallel substitutions, which renders them unsuitable for use in constructing phylogenies of plants from widely different families.[44] This difficulty is compounded because of the different mutation rates in these two proteins; plastocyanin changes more rapidly than cytochrome *c* (i.e. is less conservative) and is therefore useful at lower taxonomic levels (infra- rather than supra-familial). This might at first seem an advantage, but it is essential to be able to compare data from at least two different proteins before the data can be deemed reliable. In practice, because of the different rates of mutation, the data from cytochrome *c* and plastocyanin are not comparable, and indeed the results obtained from them for the same groups of plants are often found to conflict. Different characters of all sorts, including proteins, should be expected to evolve at different rates and in different directions, so that it is not surprising to a taxonomist to learn that different proteins (electrophoresis or sequencing) or different organs (serology) yield different indications as to relationship. The answer lies in the pooling of evidence from a wide range of proteins, preferably studied by different techniques (rather than in the total reliance upon the sequencing of a single protein), and if possible in the interpretation of the protein sequences in the light of DNA sequences.

The total amount of nuclear DNA is usually highly constant and can be useful taxonomically at levels from species to super-order.[471] In angiosperms this amount varies at the diploid level from 2×10^8 to 10^{11} base-pairs, or 0.5 to 254.8pg (i.e. a range ratio of 1:510) giving plenty of scope for variation.[28,29] Actually only about 5–10% of this is expressed genetically (transcribed and translated), and this is present as single-copy DNA sequences. Most DNA (at least 90%) consists of multiple-copy or repetitive DNA. Non-vascular plants generally have less total DNA and a lower proportion of repetitive DNA. Although repetitive DNA sequences generally lack known genic functions they are almost certainly not functionless, as they are interspersed with the single-copy DNA in a constant pattern which can be of taxonomic value.[471]

Early work attempting to utilize nucleic acid data in plant taxonomy involved methods such as *total DNA/DNA hybridization* in solution. In brief, total DNA extracted from one organism was treated to convert it to a single-stranded polynucleotide chain, and the amount of re-association (annealing) with similarly treated DNA from another taxon that occurred on mixing the two under specific conditions was taken as a measure of similarity (homology) of the nucleotide sequences. Useful results were obtained with some micro-organisms, but the existence of highly repetitive DNA sequences and three distinct DNA complements (chromosomes, mitochondria, plastids) rendered data from higher plants highly ambiguous.

The main advances in recent years have been the ability to deal with separate fractions of the DNA complement rather than having to cope with total DNA, the ability of break DNA sequences at highly specific points and to separate and characterize the fragments, the ability to bulk-up ('clone') the DNA of specific genes (via messenger-RNA) or larger fragments (direct from DNA) for subsequent analysis or use, and the development of routine sequencing methods for nucleic acids. A useful survey of data obtained from nuclear DNA and of their applications to plant taxonomy are given by Ehrendorfer.[113]

Restriction endonucleases are enzymes that break ('cleave') DNA at extremely specific points; many are known, each cleaving the DNA at a different recognition-site (nucleotide sequence) and therefore producing different, highly characteristic *restriction fragments* of DNA. These fragments can be separated by gel electrophoresis (Fig. 4.6A), and further characterized by hybridization with radioactive DNA or RNA 'probes' of known provenance. In this technique the separated restriction fragments in the electrophoresis gel are rendered single-stranded and then blotted on to a nitrocellulose filter in their same positions; hybridization with the radioactive single-stranded probe is carried out on this filter (Fig. 4.6B). The intensity of radioactivity, as shown on a photographic autoradiograph, in the restriction fragment bands is a measure of the degree of hybridization, which is in turn a measure of the similarity of the sequences of the hybridizing portions of DNA.

Nowadays, rather than total DNA being investigated, more often specific fractions are targeted in the belief that this will produce more comparable

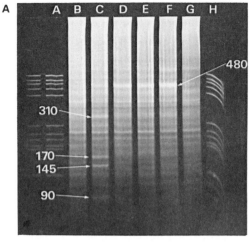

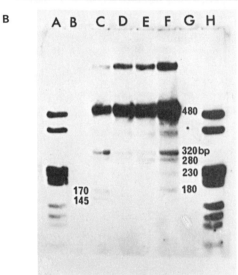

Fig. 4.6 Restriction fragments of DNA: **A**, DNA of six species of *Secale* (tracks B–G) cleaved with a restriction endonuclease; **B**, hybridization with radioactive DNA probes of gel-separated restriction fragments of the same six *Secale* species, the separated fragments having first been blotted from the gel onto a nitrocellulose filter. Taken from Jones and Flavell[238].

data (cf. use of cytochrome *c* in protein work). Good results have been obtained with, for example, single-copy DNA in *Atriplex*,[24] various defined repeated sequences in *Secale*,[238] ribosomal-RNA genes in legumes,[243] and chloroplast-DNA in potato[212] and *Clarkia*.[434] Data from work such as this have value at all taxonomic levels. At the lowest levels individual genotypes can be characterized, with obvious applications in population biology and in cultivar delimitation. One frequently used technique is to determine the length of defined restriction fragments in different populations (or species).

This will measure changes due to deletions, insertions, or even different nucleotide bases, and the pattern of distribution of such **restriction fragment length polymorphisms** (RFLPs) can be used as an assessment of the pattern of evolution. The search is also on for a particular DNA region in plants equivalent to that discovered by Jeffreys[230] permitting 'genetic fingerprinting' of human individuals. At the higher levels (genus to super-order) the results are most likely to be of value in phylogenetic studies. Although it is known that some DNA is highly conservative (resistant to change) and some evolves relatively rapidly, it seems possible that on average rates of evolution have been and are fairly constant, leading to the idea that any defined fraction of DNA is a 'molecular clock' that can be used to date particular phylogenetic divergences. Work along these lines is, to date, more advanced with birds[99] than with any plants, but encouraging results have been obtained in *Clarkia*.[434]

DNA/RNA hybridization, where RNA is hybridized on to complementary DNA of related plants, can also produce valuable information. Mabry[267] investigated the ribosomal-RNA of members of the order Caryophyllales by these methods, and concluded that the Caryophyllaceae (which contain anthocyanins instead of betalains) are quite close to the betalain-containing families, but not as close as the latter are to each other.

Nucleotide sequencing has recently become available as a routine (if tedious) technique, though it remains to be proved that it is of greater taxonomic value than assessment of similarity by hybridization. Kossel *et al.*[252] investigated ribosomal-RNA genes of certain bacteria and of chloroplasts from various plants ranging from unicellular algae to angiosperms, and found that valuable phylogenetic data can be obtained from such comparisons.

5

Chromosomal information

Chromosomal data may be viewed in two distinct ways when utilized for the purposes of classification. In one sense they are strictly anatomical information—the number of chromosomes is as important (no more or less on average) as the number of carpels, and the morphology or type of chromosomes is to be considered in the same way as the shape of leaves or petals or the type of phenolic compounds present. On the other hand they also constitute a special type of information, for chromosomal number and homology largely determine pairing behaviour at meiosis, which in part governs the level of fertility of hybrids and hence the breeding behaviour and pattern of variation of populations. Both these views are valid, and both need to be borne in mind when interpreting chromosomal data; clearly the second view assumes greater importance in biosystematic (phylogenetic) studies, but for taxonomic (phenetic) purposes the reverse may be true. There is, however, little justification for the notion that chromosomal characters are special or more important because the chromosomes carry the genes, which contain the genetic information expressed in the phenotype. In other words, the arguments used concerning the special nature of DNA sequences as taxonomic data largely seem not to apply also at the level of the chromosomes. However, the distinction between the molecular and anatomical levels of the chromosomes is becoming increasingly blurred, and there is some evidence that aspects of DNA organization above the level of base-sequence are of fundamental evolutionary significance; these aspects can be detected by modern techniques of chromosome banding and *in situ* hybridization (q.v.).

The subject matter of this chapter, often known as cytotaxonomy, can be usefully considered under three headings: chromosome number, structure and behaviour. An excellent account of these topics and their many ramifications in plants is given by Stebbins,[420] and other elementary cytological background can be obtained from textbooks such as that by White.[473]

Chromosome number

It has been realized from the early years of this century that, in general, the number of chromosomes in each cell (the ***chromosome number***) of all the individuals of a single species is constant. Moreover, except for simple multiples of that number, the more closely related species are, the more likely they are to have the same chromosome number; and the more distantly related, the more likely they are to have a different number. This relative conservativeness renders chromosome number an important and much-used taxonomic character, and it is, in fact, just about the only biosystematic evidence which is *consistently* recorded in standard Floras and the like. Usually the information is provided in the form of the diploid number ($2n$). In reporting new chromosome counts it is customary to quote the ***diploid number*** ($2n$) when the count is based on mitosis in sporophytic tissue and the ***haploid number*** (n) when it is based on mitosis in gametophytic material or on meiosis. In somatic tissue (haploid or diploid) counts are usually made on rapidly and synchronously dividing cells, for example meristems, embryos and young sporogenous tissue.

The taxonomic importance attached to chromosome numbers is evident from the wide range of compilations listing published counts. After many incomplete attempts, an extremely thorough list of all chromosome numbers of angiosperms published up to the end of 1967 appeared.[42] This is kept reasonably up to date by the *Index to Plant Chromosome Numbers* issued periodically under the auspices of a panel of plant biosystematists. This latter work, which actually commenced covering the year 1956, includes all green plants and fungi, but leaves a gap in the literature concerning pre-1956 counts of non-angiospermous plants. This gap is filled for the bryophytes by the work of Fritsch[140] (see Newton[308,309] for partial update), and less satisfactorily for the pteridophytes by Löve *et al.*[265] and for the algae by Godward.[158] Counts are now available for about 8% of bryophytes, 20% of pteridophytes and 15–20% of angiosperms, although many of the counts cannot be considered absolutely reliable and in rich floras, especially in the tropics, the percentages are much lower. Clearly, there is a great need for more chromosome counts, both of uncounted and of insufficiently covered species, and it is most important that voucher specimens of the plants used for cytological study are deposited in herbaria where their identity and characteristics can be checked by future workers.

In studying the above lists it quickly becomes evident that closely related species (say within one genus) often differ in chromosome number, the most frequent variations being based upon the phenomenon of ***polyploidy***. In the genus *Festuca*, for example, there are species with $2n = 14, 28, 42, 56$ and 70; such species are known as ***diploids***, ***tetraploids***, ***hexaploids***, ***octoploids*** and ***decaploids*** respectively. Clearly these numbers are based upon 7, the gametophytic chromosome number of the diploid species. This number is known as the ***base-number*** or ***basic chromosome number*** (x), which may be looked upon as the crudest manifestation of the ***genome***, the basic set of genetic information carried by the plant. In a diploid species $x = n$, but in a polyploid species n is a multiple of x; hence, in the case of the hexaploid fescue mentioned above, $2n = 6x = 42$ (or $n = 3x = 21$). Polyploidy is extremely widespread in plants and has been a major feature in plant evolution. Estimates of the proportion of species of angiosperms which are polyploid vary from about 30% to 70%, in marked contrast to the rarity of polyploidy in animals. In pteridophytes polyploidy is common, but in bryophytes it is of uncertain frequency because there is argument as to what constitutes the sporophytic diploid level. In hepatics, for example, 87% of species have $2n = 16, 18$ or 20, but there is a difference of opinion as to whether this represents the diploid or tetraploid level.[308,309] Unfortunately the occurrence of polyploidy has led to a dual meaning of the terms diploid ($2x$ or $2n$) and haploid (x or n) which must be understood from the context (although in the latter case the terms *monohaploid* and *polyhaploid* are often used to differentiate between '$n = x$' and 'n = multiples of x' situations respectively). Many taxonomic (and other) aspects of polyploidy are discussed in the volume edited by Lewis.[260]

The base-number is easy to determine in cases such as that of *Festuca* above, but sometimes it can only be guessed at or, at best, deduced. In the genus *Pandanus*, for example, all species so far counted have $2n = 60$. The base-number could be 5, 6, 10, 15 or 30. The fact that the only other count in the family Pandanaceae (for a species of *Freycinetia*) is $2n = 30$ suggests that the base-number for *Pandanus* is 5 or 15, not 6, 10 or 30, because it is presumed that sporophytic chromosome numbers represent even numbers of chromosome sets. (One recent count of $2n = 50$ in the family has complicated this story, but the new number needs confirmation.) Problems such as these are often intractable because the ancestral diploid ($2x$) species are now extinct; the extant polyploids are often known as *palaeopolyploids*, having evidently arisen long ago.

Groups of organisms in which there is a range of chromosome numbers representing different degrees of polyploidy (*ploidy levels*) are known as *polyploid series*, for example the *Festuca* species mentioned above. Where the inter-relationships between the different ploidy levels and between different taxa at each ploidy level are complicated it is more often referred to as a *polyploid complex*. Good examples of the latter among the British flora are afforded by the *Cardamine pratensis* complex and the *Valeriana officinalis* complex: in both these complexes specific delimitation is extremely problematical since the ploidy levels are not represented by completely distinctive sets of morphological characters, and many workers regard each complex as a single polymorphic species.

It is possible to comprehend the structure of such complexes only by studying their modes of origin. Polyploidy is known to arise by either somatic or meiotic processes. In the former case the chromosome number of a cell becomes doubled by mitotic chromosomal division not being accompanied by cell division. This cell, if it is part of a young embryo or meristem, can soon develop into a substantial piece of tissue which gives rise to a plant or shoot which produces flowers with diploid microspores and megaspores, and hence tetraploid seeds. In the latter case, a diploid plant produces diploid microspores and megaspores owing to pre-meiotic doubling or to non-reduction at meiosis, tetraploid (or triploid) seeds again being the result. If the diploid plants referred to in these two methods are distinct species, whose genomes can be represented as AA, then the tetraploids derived from them will be *autopolyploids* or *autoploids* (autotetraploids, designated AAAA). But if the diploid plant was a hybrid between two plants with unlike genomes (AB), the derived tetraploid will be an *allopolyploid* or *alloploid* (allotetraploid, AABB). In many cases the two genomes are neither quite alike (*homologous*) nor utterly unlike (*non-homologous*), but somewhere between (*homoeologous*, designated AAA'A'). Such cases are known as *segmental allopolyploids*, and should be looked upon as providing every intermediate situation between extreme autopolyploids and extreme allopolyploids. It is possible to distinguish an extreme allopolyploid from a diploid only by reference to other related species. For example, a species with $2n = 20$ could represent a diploid

($x = 10$) or a tetraploid ($x = 5$). If that species has related species with $2n = 10$ the latter situation is presumably indicated, but nevertheless the species (AABB) might well behave exactly as though it were a diploid. For this reason such extreme allopolyploids are known as **amphidiploids**. Another sort of intermediate can exist at the hexaploid (or higher) level, where one genome is represented twice and one four times (**autoallopolyploid**, AABBBB).

In the case of allopolyploids the genomes of the parental diploids may have the same base number (**monobasic polyploidy**), for example in *Festuca* (mentioned above), or the polyploid may incorporate genomes of different base-numbers (**dibasic polyploidy**), for example the three species *Spartina maritima* $2n = 60$, *S. alterniflora* $2n = 62$ and *S. anglica* $2n = 122$. In such situations the derived allopolyploid possesses a new base-number; in *Spartina* the new base-number $x = 61$ has originated from parents with base-numbers of $x = 30$ and $x = 31$.

Quite often a polyploid complex consists of a series of distinct diploid species which have hybridized and become polyploid to produce a range of tetraploid, hexaploid and sometimes higher levels of ploidy (Fig 5.1). Because of the recombination of genomes the distinction between the taxa which exists at the diploid level often becomes blurred at the tetraploid and higher levels, and distinct taxa cannot be recognized. Such a situation is described as a **polyploid pillar complex** and examples are to be seen in the *Dactylis glomerata* and *Juncus bufonius* complexes. The names *D. glomerata* and *J. bufonius* strictly refer to the tetraploid and hexaploid–octoploid representatives of their complexes respectively; in both cases a number of diploid species can be segregated from them.

The type of polyploidy referred to hitherto is known as **euploidy**, but **aneuploidy**, where chromosome numbers vary not in multiples of the basic set but in single or few chromosomes only, is also common. In the genus *Vicia* there are chromosome numbers of $2n = 10$, 12, 14, 24 and 28, obviously consisting of aneuploids clustered around the diploid and tetraploid levels, and in *Crepis* $2n = 6$, 8, 10, 12, 14, 16, 18, 22, 24, 42, 44, 66, 88 (and some other numbers), with less obvious clustering. Where the numbers bear no obvious relation to a particular series of ploidy levels the term **dysploidy** may be used. Many mechanisms whereby single chromosomes are lost or gained are now known. Whereas the gain of a chromosome is equally feasible at any ploidy level, loss of a chromosome at the diploid level is usually lethal, although it can be tolerated more readily at the higher levels because of the buffering effect of the multiple genomes. On the other hand a reduction in chromosome number at the diploid level can be accomplished with very little loss of chromosomal material, if previously the lost chromosome had become extremely reduced in size by very unequal reciprocal translocations. A diploid with one extra chromosome (i.e. that chromosome represented three times) is known as a **trisomic**, and one with one chromosome missing a **monosomic**; normal diploids are **disomics**.

The known extremes of range in the chromosome number of plants appear to be $2n = 4$ (in the dicotyledon *Haplopappus gracilis* and in some algae) and $2n = 1\,260$ (in the fern *Ophioglossum reticulatum*). If we are correct in interpreting the base-number of *Ophioglossum* as $x = 15$, then *O. reticulatum*

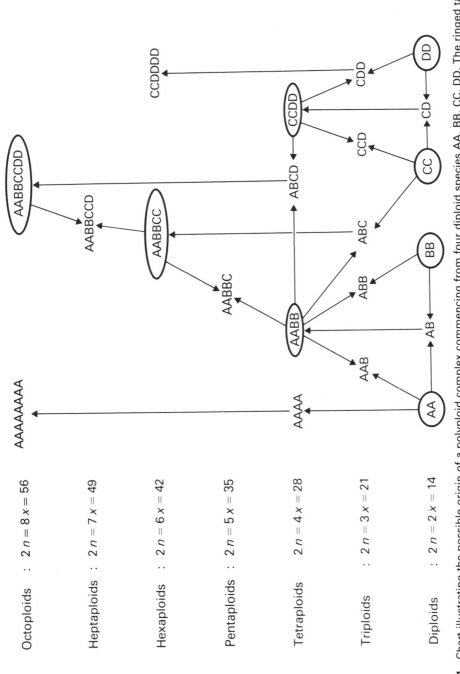

Fig. 5.1 Chart illustrating the possible origin of a polyploid complex commencing from four diploid species AA, BB, CC, DD. The ringed taxa have each genome represented twice and might be expected to exhibit full fertility, while the other taxa might be expected to exhibit varying degrees of sterility. The chromosome numbers on the left are those which would result if the whole complex had the base-number of $x = 7$.

Octoploids : $2n = 8x = 56$

Heptaploids : $2n = 7x = 49$

Hexaploids : $2n = 6x = 42$

Pentaploids : $2n = 5x = 35$

Tetraploids : $2n = 4x = 28$

Triploids : $2n = 3x = 21$

Diploids : $2n = 2x = 14$

is an 84-ploid species. The highest recorded number in an angiosperm is $2n = 264$ (?38-ploid), and in a bryophyte $2n = 132$ (?22-ploid).

Since chromosome numbers have changed in evolution through many diverse avenues, it is obvious that comparisons are valid only within certain rather narrowly defined limits. Thus the identity of chromosome number of *Festuca ovina* and *Vicia sepium* ($2n = 14$), or of *Haplopappus gracilis* and *Spirogyra cylindrica* ($2n = 4$), or of *Pyrola minor* and *Homo sapiens* ($2n = 46$), are purely coincidental. Comparisons between species within one genus are presumably meaningful, but in intermediate cases (as with the example of the sulphur compounds mentioned on pp. 90–91) one is left to reason, with the chances of a wrong decision being made.

So far chromosome number has been discussed mainly as though it were a characteristic of only species, but this is far from the truth. There are probably no families of appreciable size in which there is only a single chromosome number, although some, such as Pinaceae, nearly all with $2n = 24$, and a number of bryophyte families, come close to it. A valuable review of chromosome numbers at the family level, covering the whole of the angiosperms, is provided by Raven.[342] His main conclusion is that the original base-number for the angiosperms is $x = 7$, and that comparisons at the family level are valid only when the base-number (rather than the chromosome number) is used. In Cronquist's classification, all but one subclass (Caryophyllidae, $x = 9$) of the two classes and ten subclasses into which the angiosperms are divided have the predominant base-number $x = 7$. Moreover, the presence of numerous diploids ($2x$) in all the subclasses leads one to the conclusion that, despite the very widespread and frequent occurrence of polyploidy, the major lines of evolution of angiosperms have been at the diploid level. The same base-number does not seem to be typical of other groups of plants; for example, in liverworts the most common haploid (?base) numbers are 8, 9 and 10.[308,309]

Chromosome numbers have frequently proved useful within the family at tribal and generic levels. The case of Ranunculaceae, mostly with $x = 8$ but some genera with $x = 7$ and segregated into a separate tribe, is mentioned on p. 73. In fact one other genus has $x = 10$ and another $x = 13$, and in some current classifications these two are placed in separate tribes or even families. In the Poaceae the different subfamilies, tribes and genera can be characterized to varying extents by their base-numbers. For example, the allegedly primitive subfamily Bambusoideae has $x = 12$, while the familiar North Temperate subfamily Pooideae has mostly $x = 7$. But within the latter certain tribes (e.g. Glycerieae, $x = 10$) or genera (*Anthoxanthum*, $x = 5$) deviate consistently and illustrate the *relative* nature of the concept of base-number. In the Rosaceae the subfamily Pomoideae (apples, hawthorns, etc.) has $x = 17$, whereas in other subfamilies $x = 7$, 8 or 9. This suggests that the former are either polyploid hybrids between taxa with $x = 8$ and $x = 9$, or are polyploids of taxa with $x = 9$ with the loss of one chromosome; in either case they are an excellent example of the evolution of a new base-number via palaeopolyploidy.

Interspecific variation in chromosome numbers has proved to be one of the richest sources of cytological data of value to taxonomists. At this level there is usually a fairly obvious single base-number, from which the variations in chromosome number have been derived to produce aneuploids and polyploids. Three examples have been chosen out of the many hundreds available.

In the grass genus *Vulpia* diploids ($2n = 14$), tetraploids ($2n = 28$) and hexaploids ($2n = 42$) occur. The genus is divided into five sections of which three contain only diploids, one (*Monachne*) diploids and tetraploids, and the other (*Vulpia*) all three levels of ploidy. In section *Monachne* there are only three species, two diploids and a tetraploid, and one of the diploids and the tetraploid resemble each other very closely, so that in the past they have often been considered a single species. The diploid and tetraploid do, however, differ consistently in the pattern of ovary pubescence and, since they also show different (albeit overlapping) ecological preferences and geographical distributions, it is considered better to recognize them as distinct species.[412] In section *Vulpia* there are three diploids, two tetraploids and five hexaploids. The two tetraploids are very similar but occupy different geographical areas; since their morphological differences are only quantitative and scarcely absolute, they are, in reality, geographical races and seem best treated as subspecies of a single species.[411]

It is presumed that the tetraploid and hexaploid species of *Vulpia* arose from diploid progenitors, although the precise ancestral history of them has so far defied understanding. In some situations, however, one can be more sure of the interspecific relationships. A group of weedy *Senecio* species, for example, includes the diploid *S. squalidus* ($2n = 20$), the tetraploid *S. vulgaris* ($2n = 40$), and the hexaploid *S. cambrensis* ($2n = 60$). Since the last was discovered within living memory in an area of Wales occupied by the two other species, and is intermediate between them in morphology (Fig. 5.2), and since sterile triploid hybrids ($2n = 30$) between the two former species have been found in a few places in Britain, it seems clear that *S. cambrensis* is

1 cm

Fig. 5.2 Photographs of flower-heads of *Senecio vulgaris* (**V**), *S. squalidus* (**S**) and their amphidiploid derivative *S. cambrensis* (**C**).

an allohexaploid derivative of the other two species. Moreover, plants closely resembling it have been synthesized from its putative parents. The recognition of *S. cambrensis* as a separate species is thus justified on the basis of many comparable examples in other genera.

The genus *Vicia* contains species with $2n = 10$, 12, 14, 24 and 28, as mentioned above; of these, $2n = 12$ and 14 are very common but the other numbers are much less so. It seems very likely that the tetraploids with $2n = 24$ are derived from diploid ancestors with $2n = 12$, and those with $2n = 28$ from ancestors with $2n = 14$. Generally, the species are characterized by a single chromosome number, which thus serves as a useful additional taxonomic character.

At what may be termed 'around' the species level there is frequently a good deal of variation in chromosome number. Very often the decision as to whether this represents interspecific or intraspecific variation depends upon the extent to which chromosome number is used in determining specific limits. Since plants with different chromosome numbers are usually effectively genetically isolated (even if they hybridize their progeny are likely to be infertile), it is often claimed that each species should be represented by a single chromosome number. Á. Löve[264] has expressed this view in its extreme form, together with the notion that differences in chromosome *base-number* should not be tolerated within a single *genus*. By use of these rigid criteria many new genera have been described and many species segregated when the use of morphological data alone would not suggest their separate recognition. It is true that the great majority of species separated largely on cytological evidence were earlier recognized as taxonomic entities on morphological grounds,[263] but the differences are often very slight and had often led to recognition at only the infraspecific level. The redefinition of genera and species by means of chromosome number alone, as described above, cannot be justified.

Many good examples illustrating a range of morphological divergence associated with differences in ploidy level are found in the ferns. The three cytological races (diploid, tetraploid, hexaploid) of *Polypodium vulgare* have almost equally been regarded as species or subspecies; the diploid and tetraploid races of *Asplenium trichomanes* have usually been considered only as subspecies; and many of the various ploidy levels of *Dryopteris filix-mas* are nowadays almost universally recognized at the specific level. Such examples can be paralleled in other groups, for example bryophytes and flowering plants. Thus the closely related moss species *Mnium punctatum* and *M. pseudopunctatum* are haploid and diploid respectively, while the haploid and diploid races of *Pellia epiphylla* are rarely separated taxonomically (though the latter is *sometimes* segregated as *P. borealis*). The detailed work of Newton[309] has shown that the chromosome number (and structure) of mosses offers much of taxonomic and evolutionary significance. In the flowering plants the diploid and tetraploid watercresses are usually separated as *Nasturtium officinale* and *N. microphyllum*, the different ploidy levels of *Galium palustre* are variously recognized as species or subspecies, and the diploid and tetraploid races of *Ranunculus ficaria* are considered either subspecies or varieties. These examples illustrate that, while there is a difference of opinion among taxonomists, it is a difference, in the main, only in detail and there is a consensus that morphological criteria should not be totally overruled by chromosome numbers.

The terms *cryptic polyploidy* and *semi-cryptic polyploidy* can be usefully employed to cover cases where there is no or little external manifestation of differences in ploidy level. Despite some claims to the contrary, there certainly appear to be species in which there exists truly cryptic polyploidy or aneuploidy; *Alisma plantago-aquatica*, *Mimulus guttatus*, *Galium verum* and the liverwort *Lophocolea bidentata* are examples. This is not surprising, for it is well known that even different tissues within one plant can vary in chromosome number (*mixoploidy*), particularly in relation to the differentiation of organs. Autopolyploids in general show few differences from the diploids whence they arose (except that they are often sterile), and are rarely recognized as taxonomic entities; the same is true of recently derived aneuploids (trisomics, etc.).

Taxonomic problems often arise because morphological differences between taxa with different chromosome numbers exist but are not absolute, i.e. the chromosome number can usually be predicted from a morphological study, but not always. Such is the case in the diploid and tetraploid *Ranunculus ficaria* mentioned above, and in various segregates of the *Vicia sativa* aggregate. In this group $2n = 10$, 12 or 14 and, while some morphologically defined taxa are characterized by a single chromosome number, others appear not to be.[209,210] The taxa are recognized as either species or subspecies by different taxonomists. Populations or infraspecific taxa differing in chromosome number (or chromosome morphology) are often known as *cytotypes*.[446]

Chromosome structure

The most commonly utilized aspect of chromosome structure is the position of *centromeres*, i.e. the arm-length ratio of each chromosome in the genome. Sometimes it is sufficient only to recognize the distinction between chromosomes with centromeres near the middle (*metacentrics*) and those with centromeres near one end (*acrocentrics*), but in other cases a more precise definition is required; in addition *telocentrics*, with a truly terminal centromere, occur (Fig. 5.3). Several schemes different from the one shown in Fig. 5.3 have been proposed in recent years.

Overall size of chromosomes can be measured in absolute terms, although this carries great problems of standardization which are not altogether solved even by the measurement of DNA content (see p. 106). More often a relative size measurement is satisfactory; usually a distinctive chromosome is taken as standard and the lengths of other chromosomes in the genome calculated relative to it, or each arm-length is calculated as a proportion of the total genome length.

A third useful aspect of chromosomal gross morphology is the position of *secondary constrictions*, which delimit the occurrence of *satellites*. The presence of satellites is important not only for their intrinsic taxonomic value but also because they can be confused (especially in poor preparations or where the secondary constriction is greatly elongated) with additional chromosomes. Examples of this can be seen in *Calystegia* (erroneous counts of

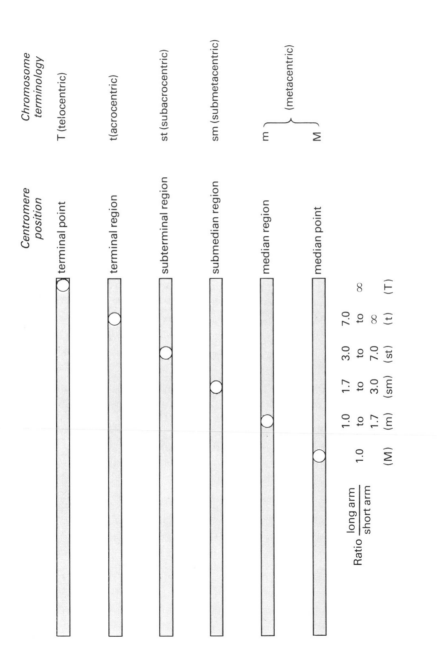

Fig. 5.3 Diagram illustrating chromosomal terminology according to centromere position, as defined by Levan et al.[258]

$2n = 24$ instead of $2n = 22$)[404] and *Ononis* (errors of $2n = 32$ for $2n = 30$).[298] The recording of satellites is problematical because they are often very variable in appearance, at some times being very conspicuous and at others indiscernible, and it is not always possible to obtain consistent results.

The appearance of the basic chromosome set (genome) under the light microscope is known as the **karyotype**. For example, one can talk of the karyotype of *Calystegia* ($2n = 22$) as consisting of eight short metacentrics, two short submetacentrics and one short metacentric with in addition a well-defined secondary constriction and satellite. For more accurate purposes arm-ratios can be used. Karyotypes are usually calculated from the means of a substantial number of observations, and often they are represented diagrammatically as ideograms or **karyograms** (Fig. 5.4). Some workers have devised formulae to summarize karyological data concisely, but none of these methods has been generally adopted. Nevertheless, the acceptance of a shorthand formula to designate the karyotype is surely a prerequisite for the more widespread use of cytotaxonomic data, and would probably lead to its routine inclusion in Floras, etc. The karyotype of *Calystegia* (above), using the system devised by Fernandes,[133] is $2n = 22 = 16pp + 4mp + 2mp'$.

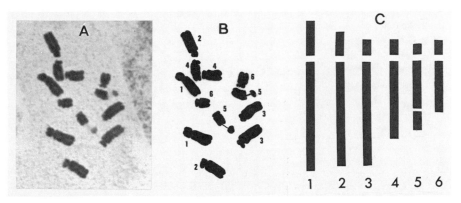

Fig. 5.4 Construction of a karyotype ideogram; **A**, photograph of root-tip mitosis; **B**, drawing of same; **C**, haploid ideogram calculated from the means of ten such drawings. The karyotype illustrated is a strain of *Vicia sativa* with $2n = 12$. Prepared, photographed and drawn by J. P. Bailey.

These aspects of chromosome structure, together with chromosome size and number, have been found extremely useful at all levels of the taxonomic hierarchy[225] (Fig. 5.5). The importance of chromosome size has already been mentioned in relation to the tribes of Ranunculaceae. Perhaps its best known use is in the monocotyledonous genera *Yucca, Agave* and relatives. These are very large plants with long, strong leaves borne in a tight rosette; they live for a considerable number of years in the vegetative state and then bear an inflorescence. After the seeds are shed the plant either dies or spends another period of some years in the vegetative state before flowering again. *Yucca* and some others have flowers with a superior ovary, while *Agave* and others have an inferior ovary; for this reason the former were previously placed in the

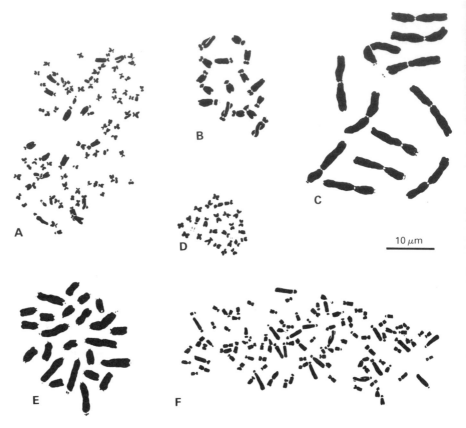

Fig. 5.5 Karyotypes of various members of the family Commelinaceae, taken from Jones and Jopling[240] and showing great range in chromosome number and morphology: **A**, *Tradescantia blossfeldiana* (2*n* = 90); **B**, *Gibasis* sp. aff. *geniculata* (2*n* = 16); **C**, *Tradescantia paludosa* (2*n* = 12); **D**, *Ballya zebrina* (2*n* = 26); **E**, *Tradescantia micrantha* (2*n* = 24); **F**, *Tradescantia fluminensis* (2*n* = 108).

Liliaceae and the latter in the Amaryllidaceae. Because of the great overall similarity in these genera they are now usually segregated from their respective families and placed in the Agavaceae. This classification is strongly supported by evidence obtained in the 1930s, when it was found that the Agavaceae are characterized by a very distinctive bimodal karyotype consisting of five large chromosomes and 25 small ones.[276] Such striking examples are very rare, but smaller differences in chromosome morphology have very often proved conclusive in taxonomic debate. The karyotype is, in effect, a more precise manifestation of the genome than chromosome number alone, and hence its use is greater when investigating genomic homologies. This is nowhere better illustrated than in *Crepis*, where the use of chromosome number and morphology has solved many taxonomic problems at the generic, sectional and specific levels, and shown how many species are evolutionarily inter-related.[13]

Study of the morphology of chromosomes over a wide taxonomic range has led many workers to deduce that, within the angiosperms, a *symmetrical karyotype* (i.e. little variation between chromosomes, and each chromosome more or less metacentric) is relatively primitive and an asymmetrical one more advanced, although many reversals of this trend must have occurred. In many families, more asymmetrical karyotypes are found in plants with more specialized morphological features, for example, *Aconitum* and *Delphinium* in the Ranunculaceae. Recently Jones[239] has challenged this theory and postulated that symmetrical karyotypes originally arose from telocentric chromosomes by end-to-end fusion. The well-known processes of translocation, inversion, duplication and deficiency of various parts of chromosomes may also readily explain the rapid and large changes in chromosome structure which must have taken place in many plants.

In recent years (from 1970 onwards) our ability to distinguish morphologically between chromosomes, at least in favourable material, has been greatly enhanced by new staining techniques using Giemsa and fluorochrome dyes, which under certain controlled conditions stain chromosomes in a consistent banding pattern instead of with a uniform intensity, as is the case with the usual basic fuchsin (Feulgen reagent).[375] These extra morphological criteria have provided distinctions between chromosomes which were previously indistinguishable or scarcely distinguishable, for example, all 23 of man, all 7 of broad bean, all 21 of wheat, and have therefore enabled a very precise characterization of the genome (and hence much better bases for their close interspecific comparison). Today many different techniques involving Giemsa and fluorochrome stains are available, each giving a different pattern of banding based on the specific biochemical consequences of each method. The variants that have proved of most value in plants so far are known as *C-banding*, *G-banding*, *Q-banding* and *Hy-banding*.[143] The extent to which the actual biochemical basis is understood varies a good deal, but provided the results are repeatable they are all equally acceptable as taxonomic evidence. In some cases banding is known to be associated with regions of *heterochromatin* (as distinct from the usual euchromatin) in the chromosomes; these two types of chromatin differ in the timing of their cycle of condensation (one being out of phase with the other) and had previously been found of some use taxonomically.

There are greater difficulties in applying these techniques to plants than to animals (perhaps due to the cell-wall), but they have become standard cytological tools in plants and much useful taxonomic evidence has already been accumulated. As an example, the conclusion from work on cultivated species of *Scilla*[168] may be cited: 'The banding data support the systematic grouping proposed on a morphological basis, and provide additional evolutionary evidence' (particularly on the origin of the cultivated taxa). In *Anacyclus* differences in the banding pattern of the various species are well marked (Fig. 5.6), although, using standard techniques, the karyotypes appear very similar.[374]

C-banding can be very useful in indicating the position of centromeres in those cases (notably bryophytes) where the latter do not or scarcely show up by conventional staining. Similarly, the technique of *silver-staining* has recently (1975) been developed to highlight specifically nucleolus-organizer

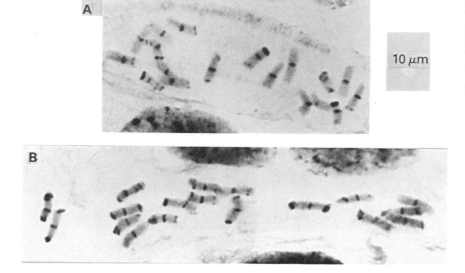

Fig. 5.6 Photographs of Giemsa-stained chromosomes at metaphase of: **A**, *Anacyclus clavatus*; **B**, *A. coronatus*, taken from Schweizer and Ehrendorfer.[374] Both species have 2n = 18 but in *A. coronatus* many more of the chromosomes have distally staining areas.

regions which, as mentioned earlier, are frequently uncertain in appearance with normal techniques.

The technique of *in situ* **hybridization** (either DNA/DNA or DNA/RNA) is also aimed at pinpointing specific areas in a genome. A radioactive single-stranded DNA or RNA probe is introduced on to a normal mitotic cell-squash in which the DNA duplex has been denatured, and the squash is then autoradiographed. Regions where the probe has hybridized *in situ* with chromosomal DNA are shown up by the presence of abundant silver grains on the radiograph (Fig. 5.7). According to the probe used, different regions, e.g. nucleolus-organizers, can be detected.

The recent convergence of biochemical, genetical and anatomical approaches to the study of chromosome organization promises to provide important new insights into plant variation and evolution, with inevitable far-reaching taxonomic consequences.[27,113,114]

Some groups of plants possess very distinctive types of chromosomes which provide important taxonomic data. For example, small chromosomes without discernible centromeres, often said to have diffuse or non-localized centromeres, occur in the Juncaceae and Cyperaceae, two monocotyledonous families which are now considered closely related, although previously placed far apart because of their quite different flower structure. Since such chromosomes do not depend upon the presence of a discrete centromere for regularity of mitotic and meiotic behaviour, their fragmentation is not necessarily deleterious and irregular chromosome numbers are characteristic of many species. In the *Luzula spicata* group, for example, plants can have 2n = 12, 14 or 24, but the total chromosomal volume is about the same in all and indicates that the higher numbers are probably derived by fragmentation

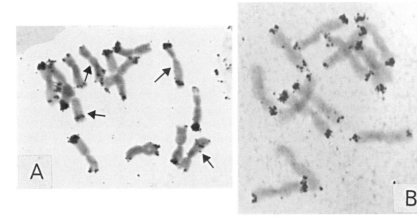

Fig. 5.7 *In situ* hybridization of metaphase preparations of *Secale* (**A**, *S. cereale*; **B**, *S. vavilovii*) with the same radioactive DNA probe, taken from Jones and Flavell[238]. Note that chromosomes with interstitial labelling (arrowed in *S. cereale*) are absent from *S. vavilovii*.

(*agmatoploidy*). In other cases different chromosome numbers occur in different cells of one root-tip (mixoploidy).

Other special sorts of chromosomes may occur together with normal ones. *Sex chromosomes* are rarely of much taxonomic value in plants as dioecism (at least in higher plants) is relatively rare and, even when it does occur (as well as in most lower plants), sex chromosomes are often not discernible. However, *Silene dioica* has an XX/XY system, *Rumex acetosa* has an XX/XYY system, and in many liverworts there is an XX/XY system where the sex chromosomes may be normal or minute in size compared with the other chromosomes (*autosomes*). The X-chromosome of *Sphaerocarpos texanus* differs in size and shape between Europe and America, but there are no associated morphological differences and a suggestion that the species be split into two has not been accepted generally. The existence and types of sex-chromosomes in bryophytes is discussed by Newton.[309]

More common are *accessory chromosomes* which are additional to the normal autosomes (*A-chromosomes*) and not homologous with them.[353] They are usually smaller than the autosomes and largely or entirely heterochromatic, and they are often not constant in occurrence. In higher plants they are known as *B-chromosomes* and an adaptive role has been demonstrated for some of these, but nearly always they have no measurable effect on the appearance of the plant and are of no taxonomic importance. In *Ranunculus ficaria* diploid plants have no B-chromosomes but 0–7 are present in tetraploids; in *Festuca ovina* there are 0–2 B-chromosomes in both diploids and tetraploids. They are designated as additional to the A-chromosomes, for example $2n = 14 + 2B$. B-chromosomes also occur in bryophytes, where there are also frequently very small chromosomes (*m-chromosomes*) that can be highly diagnostic. In some cases these seem to be constant constituents of the genome, while in others they are highly heterochromatic and of less constant occurrence, i.e. it seems that both A- and B-chromosomes can occur in the form of m-chromosomes.[309]

Chromosome behaviour

The phenomena discussed here are the pairing behaviour and the subsequent separation of chromosomes at meiosis. Not only does the regularity of pairing largely determine the fertility of a plant, but it enables a chromosome-for-chromosome comparison of the degree of homology between genomes, possibly carrying forward that comparison further than can any study of chromosome morphology. The study of chromosome pairing is one of the major means of investigation in *cytogenetics*, the study of the role of the chromosomes in heredity.

Some taxonomic information can be gained from the study of the mechanism of meiosis itself. For example meiosis in some Juncaceae and Cyperaceae, which have small chromosomes with unlocalized centromeres, is considered to be 'inverted', i.e. the equational division precedes the reductional division instead of *vice versa*. Judging from the occurrence of inverted meiosis in animals, this phenomenon is not always correlated with unlocalized centromeres, and it therefore provides further evidence for the close relationship of the above two families.

Many peculiarities of meiotic behaviour owe their origin to the existence of heterozygosity, so that at meiosis one is observing the pairing of unlike genomes. The differences which can exist have often arisen by duplications, deficiencies, inversions or translocations of chromosomal material, and the meiotic configurations usually provide evidence of the precise nature of these rearrangements.[433] Some species are known to be consistently heterozygous for certain translocations, which give rise to multivalent formation at meiosis. In the genus *Oenothera* all species are diploid with $2n = 14$. Many of these exhibit a normal meiosis, but in the subgenus *Oenothera* the species are heterozygous for translocations involving varying numbers of chromosomes, so that multivalents of varying sizes arise at meiosis.[68] In the *O. biennis* group, for example, *O. biennis* itself exhibits a ring of six and ring of eight chromosomes, *O. erythrosepala* exhibits a ring of 12 chromosomes and a bivalent, and *O. strigosa* exhibits a single ring of 14 chromosomes.

But the situation in *Oenothera* is very exceptional. Usually the occurrence of abnormalities such as this is not consistent, but represents the product of chance hybridization between two plants with genomes which are sufficiently unlike to cause mechanical problems in pairing. Such chance hybrids are therefore valuable in assessing genomic homologies, and where they do not occur in nature they can often be synthesized artificially (see Chapter 6).

When the genomic differences have arisen relatively recently they are not likely to be associated with any exomorphic differences, and are mostly of a relatively straightforward nature, for example an additional chromosome, which in a diploid might form a trivalent with its two homologues, or a reciprocal translocation, which in a heterozygote would give rise to a quadrivalent at meiosis. Many species are known in which such differences have arisen in different populations, as has been shown by the study of meiosis in artificial inter-population hybrids. In other cases such changes, coupled with geographical isolation, have led to a greater degree of divergence and the evolution of new species. That famous pair of species, *Crepis neglecta* ($2n = 8$) and *C. fuliginosa* ($2n = 6$), from S.E. Europe, is just one of many known examples of this. The extra pair of chromosomes in *C. neglecta* carries no important genes except on the distal part of one arm; translocation of this part on to another chromosome resulted in a 'genetically inert' chromosome whose accidental loss led to a viable new taxon with one fewer chromosome pair. This sequence of events can be postulated with a high degree of certainty from a study of meiosis in the interspecific hybrids ($2n = 7$), in which the pairing behaviour of individual segments of chromosomes may be followed.

Greater degrees of genomic non-homology result in non-pairing at meiosis (**asynapsis**), or loose pairing without the formation of chiasmata so that the bivalents fall apart before metaphase (**desynapsis**). In extreme cases the whole genome fails to pair; in other situations varying proportions of the chromosomes form bivalents. The degree of pairing is broadly proportional to the level of homology of the genomes, but not exactly so, since relatively simple inversions, etc., can lead to total loss of synapsis. The extent of pairing is certainly not strictly correlated with the level of fertility of a hybrid, since there are many other events besides asynapsis which may lead to sterility. Thus it follows that chromosome behaviour is only one of many factors determining the pattern of variation and taxonomic distinctness. Nevertheless, the study of chromosome pairing in diploid hybrids is a very valuable means of assessing genomic homology, which may be important in both taxonomic and evolutionary studies; such investigations constitute **genome analysis**. In reporting results of pairing studies, univalents are represented by I, bivalents by II, etc. Hence a triploid hybrid might be designated $2n = 21 = 7\text{II} + 7\text{I}$.

A traditionally fruitful use of genome analysis in plant taxonomy has been the investigation of polyploids in order to ascertain their ancestral genomes. In a genus with, say, ten diploid species, the genomic constitution of each of the species may be designated AA, BB, CC, $----$ JJ. In fact, after the study of meiosis in diploid–diploid hybrids, one might conclude that not all the diploids have different genomes; thus two species might share genome BB (homology) or the differences between two genomes might be better indicated by BB and B'B' rather than BB and CC (homoeology rather than non-homology). If there were a single tetraploid in that genus one could designate its genomes PPQQ. But, presumably, such tetraploids have been derived from diploids, and it has very often proved possible to detect the genomes involved among still existent diploid species. If triploid hybrids are made between the tetraploid and all ten deploids (PQA, PQB, etc.), the identity of genomes P and Q with one or two of A–J can often be established by chromosome pairing behaviour. If each genome contains seven chromosomes, the triploid PQA will form twenty-one univalents if A is not homologous with P or Q, but seven univalents and seven bivalents if it is homologous with either P or Q (Fig. 5.8). In practice one might not need to synthesize all ten possible hybrids, because the most likely parental genomes are frequently indicated by particular exomorphic characters of the species involved.

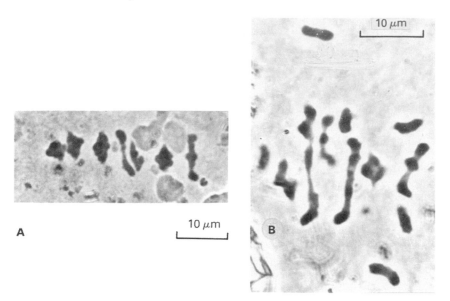

Fig. 5.8 Meiosis in: **A**, a diploid species ($2n = 14$); and **B**, a triploid hybrid ($2n = 21$) of *Vulpia*, showing seven bivalents and regular division in the former, and seven bivalents plus seven univalents and irregular division in the latter. Photographs by C. M. Barker.

By use of these methods the parentage of a great many polyploids has been established beyond any reasonable doubt. Perhaps most significant is the case of wheat, *Triticum aestivum*, a hexaploid with the designation AABBDD. The A genome is present in the diploid *Triticum monococcum*, the B genome in the diploid *Aegilops speltoides*, and the D genome in the diploid *Aegilops squarrosa*. It is important to realize, however, that except in very recently derived polyploids the statement that a genome is shared by a diploid and a polyploid does not necessarily imply (or perhaps rarely implies) that the diploid itself was ancestral to that polyploid—rather that a diploid with that genome, or one very like it, was ancestral to that polyploid. This is in fact partly the case with wheat, whose genomic constitution is nowadays more usually expressed A′A′B′B′DD. The genomes ancestral to a particular polyploid are often detected in species within different genera (as in the wheat example), but of course it might well be that at the time of formation (perhaps millions rather than thousands or hundreds of years ago) the taxa were less distinct than they are now, so that a contemporary taxonomist, had he existed, would not have placed them in different genera. Nevertheless, there are numerous cases where the close genomic homology and the completely predictable exomorphic characteristics of the diploids and polyploids lead one to believe that the latter are indeed directly descended from the former. In a small number of instances, where the polyploid has arisen in historical times, this is known to be so, for example, *Primula kewensis* from *P. floribunda* and *P. verticillata*, *Senecio cambrensis* from *S. vulgaris* and *S. squalidus*, *Spartina anglica* from *S. alterniflora* and *S. maritima*, *Tragopogon mirus* from *T. dubius* and *T. porrifolius*. At the opposite end of the spectrum the ancestry of most polyploids is quite unknown. This is usually due to lack of detailed studies, but where such investigations have proved fruitless it could be that the ancestral diploids have become extinct, or that they have diverged so much since giving rise to the polyploid that we cannot recognize them, or even if we suspect them as ancestors it is no longer possible to hybridize them with the polyploid.

In a completely fertile allopolyploid the chromosomes generally form only bivalents at meiosis. If that allopolyploid is a tetraploid, AABB, one would infer that the bivalents were A-A and B-B pairs, and that A and B were non-homologous. If, on the other hand, A and B were homologous or homoeologous, one would expect, in addition to A-A and B-B pairing, some A-B pairing and hence some quadrivalents (A-A-B-B) at meiosis, i.e. the tetraploid would be an autotetraploid or segmental allotetraploid, according to the amount of pairing. A-B pairing would also show up in the diploid hybrid (AB) from which the tetraploid arose. Thus it follows that the less pairing (i.e. the greater sterility) in the diploid hybrid, the greater the regularity of bivalent as opposed to multivalent formation (i.e. the greater fertility) in the derived tetraploid. This apparent anomaly has been well substantiated in many instances. The existence of so many intermediate situations (with some but not full pairing in the diploid, and some multivalent formation in the tetraploid) serves as a reminder that allopolyploidy and autopolyploidy are only relative terms, or extremes of a range. Such intermediates arise from the hybridization of species which share a genome in common or have similar genomes, or from the hybridization of partially differentiated races of a single species.

Unfortunately for students of flowering plant evolution genomic homology is not the only factor which determines synapsis. Solbrig[395] has listed 15 genes which are known to affect the regularity of meiosis, several of which are concerned with pairing itself. The most conclusive evidence was presented by Riley and Chapman,[362] who found that in hexaploid wheat there is a gene (on chromosome 5 of the B genome) which suppresses multivalent formation. In strains of wheat in which that chromosome, or the relevant part of it, is lost, hexaploid wheat forms some multivalents at meiosis, indicating that it is not an amphidiploid as had been supposed, but a segmental allohexaploid 'diploidized' by gene action. The partial homology of the A, B and D genomes thus revealed is also detectable in some of the primary ancestral hybrids, and in triploid polyhaploid wheat with the constitution ABD.

This discovery has had extremely far-reaching repercussions in the field of plant biosystematics. In the case of a pair of species, the one diploid (AA) and the other tetraploid (PPQQ) and both forming only bivalents at meiosis, if the triploid hybrid between them (APQ) formed equal numbers of bivalents and univalents at meiosis (Fig. 5.7), the classical deduction would be that the diploid genome (A) is identical with one of the tetraploid genomes (P or Q) and that the diploid is ancestral to the tetraploid. But Riley and Chapman's discovery provides a second possibility: that the bivalents in the triploid APQ are not A-P or A-Q, but P-Q pairs, even though in the tetraploid there was none of the latter. In that case the diploid would not be ancestral to the tetraploid. There are now many cases where apparent amphidiploids have turned out to be diploidized segmental allopolyploids, thus nullifying a great deal of biosystematic research. In order to distinguish between the two situations one must study meiosis in plants possessing only one of each type of genome present in the polyploid (PQ in the last example). Such plants may exist as wild hybrids or be obtained by artificial hybridization of the parental diploids or by the experimental production of polyhaploids from the polyploids. But most often all these means fail and no answer to the dilemma is forthcoming.

The traditional methods of genome analysis have, therefore, fallen out of favour with many workers, especially as genes suppressing multivalent formation have been shown to be widespread. In some groups, for example Poaceae, they are the rule, but in others, for example Brassicaceae and ferns, they are rare or have not yet been detected. Hence in taxa of the latter type, genome analysis by means of the study of synapsis in hybrids carries on unabated and with extremely rewarding results. Nowhere has it been better put to use than in the ferns, particularly the genera *Dryopteris*, *Polystichum*, *Polypodium* and *Asplenium*, and the ancestral relationships of species in these genera are as well known as those of any plants. In economically important groups such as the grasses (Poaceae), where the diploid hybrids or polyhaploids are not obtainable in most cases, the future of genome analysis probably lies in less direct techniques, such as the analysis of isozyme patterns or Giemsa staining patterns, or *in situ* DNA/DNA hybridization studies. The application of molecular biological methods to genome analysis seems sure to pay rich dividends in the future.

6

Information from breeding systems

The *breeding system* of a plant may be defined broadly as the mode, pattern and extent to which it interbreeds with other plants of the same or of different taxa. *Inbreeders* are plants which predominantly or wholly produce seed from self-fertilization, *outbreeders* those which produce it from cross-fertilization; there is in nature, however, every situation from one extreme to the other.

The view taken here is that the breeding system is important taxonomically for three reasons: the extent of interbreeding largely defines the pattern of variation and hence the delimitation of taxa; a knowledge of the breeding system frequently helps to understand taxonomic complexity, although often it does not solve the problems associated with it; and a study of the breeding system is often vital in unravelling evolutionary pathways.

The extent to which the breeding system determines variation patterns is a point of argument at two separate levels. Firstly, the proportion of inbreeding/outbreeding within any given species is traditionally considered to determine the variation within and between populations. In outbreeding species each population is variable but similar to nearby populations, due to the gene exchange which takes place between such populations. Inbreeding species tend to exist as relatively uniform populations which, however, often differ considerably one from another, even over quite short distances, since gene exchange between them is low or absent. These two different patterns and an intermediate one are illustrated in Fig. 6.1. This concept has been challenged recently, notably by geneticists who have discovered considerable heterozygosity (one form of intra-populational variation) in what were previously considered to be strictly inbreeding taxa. To what extent this can be

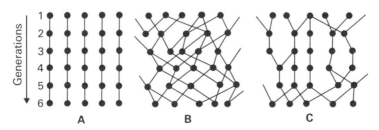

Fig. 6.1 Diagrammatic representation of obligate inbreeding (**A**), obligate outbreeding (**B**), and an intermediate situation (**C**), adapted from Heslop-Harrison.[197] The dots represent individuals, which are arranged in six horizontal rows representing six generations. The lines show the ancestry of the individuals through the six generations.

considered the remnants of variation from a time when the species was more outbreeding, or the results of mutation or of very occasional cross-pollinations, is uncertain. But it does not alter the general conclusion that more outbreeding leads to greater intra- and less inter-population variation. In plants which are definitely wholly lacking in cross-pollination, i.e. obligate apomicts, there is usually very little, and often no, demonstrable genetic variation within populations. Solbrig[396] stated that the 'breeding system is only one of the many factors regulating recombination in plants', and gave a table, taken from Grant,[163] which listed nine factors (Table 6.1). However, Solbrig was using the term breeding system much more strictly than in this book, where it is used almost to mean 'recombination' in the sense of Solbrig.

Table 6.1 Factors regulating genetic recombination in plants, taken from Solbrig.[396]

A. Factors controlling the amount of recombination per unit of time
    1. Length of generation

B. Factors controlling the amount of recombination per generation
    2. Chromosome number
    3. Frequency of crossing-over
    4. Postzygotic sterility barriers
    5. Breeding system
    6. Pollination system
    7. Dispersal potential
    8. Population size
    9. Crossability barriers and external isolating mechanisms

Secondly, the extent to which a taxon interbreeds with a different taxon is often considered a measure of the phenotypic discreteness and hence taxonomic validity of that taxon. In other words, if hybrids arise they blur the distinction of the two taxa. It is now clear that this is by no means always true. Lewis[259] claimed that 'rate of gene exchange has no taxonomic significance whatever'. He based his belief partly on the observation that autopolyploids are genetically isolated from their parental diploids yet remain phenetically very similar to them, and partly on the fact that geographically totally isolated parts of a fully interfertile taxon are usually not recognized as distinct species. Hence, he argued, taxonomic species are defined on phenetic discontinuities, not on interfertility. Nevertheless, the degree of phenetic discontinuity is itself at least partly determined by the ability or not to interbreed, and on the length of time during which barriers have existed. Raven[343] considered that gene flow (i.e. interbreeding) is almost always extremely localized, and effective population size is therefore small. Hence gene flow cannot account for the similarity of individuals in what we define as species, and 'alternative explanations' for the existence of recognizable species in nature 'must be formulated'. On the other hand Loveless and Hamrick[266] concluded that gene-flow appears to be important in distributing alleles widely within a species. It is clear that a detailed consideration of the breeding system is essential for a complete taxonomic understanding of any group.[246] A useful survey of flowering plant breeding systems has been provided by Richards,[357] and they are neatly put in a wider biosystematic context by Briggs and Walters.[53]

Ideal species

As a starting point, one may take the example of an 'ideal' species. By this is meant a taxonomically ideal species, i.e. one which poses no taxonomic problems, as it is always recognizable as a distinct entity. Thus it has no sharp discontinuities of phenotype within its spectrum of variation, and does not merge with other species.

In most instances such species are genetically isolated from (i.e. unable to interbreed with) other species and are at least partially outbreeding. These are the 'sexual, outbreeding non-hybridizing species' discussed by Stace.[407] Species of this type are very common in, for example, many Fabaceae and Apiaceae, *Sedum*, *Campanula* and *Allium*, although there are exceptions in all these examples.

Most (though not all) taxa which pose intrinsic taxonomic problems do not correspond with the above situation. Instead, either they hybridize with other taxa so that the genetical limits are wider than the morphological ones, or they exhibit breeding barriers between different elements of a morphologically recognizable taxon. These two categories are discussed separately below.

Hybridizing species

The subject of plant *hybrids* and *hybridization* has been treated in detail by Stace[405] with special reference to the British flora. In this chapter only the salient points relating to taxonomy are covered.

A hybrid has been defined as 'a zygote produced by the union of dissimilar gametes'.[95] Such an extremely broad definition is quite valid in a genetical sense, but is useless in a taxonomic one since it covers almost all animal individuals and the majority of plant individuals. A more useful definition is of a *taxonomic hybrid*—the product of breeding between distinct taxa. One can be more or less specific by limiting the context to taxa of particular levels, for example interspecific hybrid, intergeneric hybrid. Usually taxonomists, when speaking of hybrids, are referring to interspecific (or intergeneric) hybrids rather than to hybrids between taxa within one species.

The extent of hybridization

The notion that interspecific hybrids are rare is ill-founded. The only modern broad compilations of their numbers (i.e. number of species combinations) are those of Knobloch,[248,249,250] who recorded 23 675 in the angiosperms and gymnosperms and 620 in the pteridophytes. However, this was an uncritical survey, many fictitious hybrids being included, as were those only artificially produced. In terms of well-substantiated, naturally occurring hybrids the figure is much lower, but there must be a great many undetected hybrids in underworked areas, especially the tropics. Realistic figures can be obtained only from very well studied floras, such as that of the British Isles. Stace[405] recorded 626 'well substantiated interspecific' hybrids among vascular plants, plus 122 possibly correctly identified ones (as well as 227 errors) in the British Isles; since that compilation at least 100 additional combinations have been discovered. Taking into account the doubtful cases above, and a few errors in the 626, it is reasonable to assume that there are about 780 interspecific hybrids among the British vascular plant flora of about 2 500 native and alien sexual species (i.e. excluding apomictic taxa), ignoring the many combinations recorded from only outside the British Isles yet involving two British species. The figure of 2 500 represents, very approximately, 1% of the world's vascular flora, which might therefore include in the region of 78 000 different naturally occurring interspecific hybrids. Although most of these are uncommon, clearly hybridization is not a rare or abnormal phenomenon. Moreover, a great many firmly accepted species are known to be of hybrid origin. According to the levels of polyploidy considered to be exhibited by angiosperms, this applies to between 30 and 70% of species. When one considers the *potential* for hybridization, hybrids assume even greater importance. There are vastly more artificial hybrids known than natural ones; in the Orchidaceae alone about 75 000 have been synthesized. Indeed, one is forced to the conclusion that in general the ability to hybridize is the usual situation.[343] In the British flora about half the genera are monotypic, and about half the rest form at least one hybrid in the wild.

A small proportion of naturally occurring hybrids involves three or even more species. This, of course, is always a possibility when a primary (dispecific) hybrid is fertile, since it might then mate with a third species. About 31 such hybrids have been recorded from the British Isles. Again, the possibilities are much greater in the case of artificial hybridizations, and a hybrid incorporating 13 different parental species of *Salix* has been synthesized in Sweden.[310]

In the non-vascular groups much less is known concerning natural hybridization. In terms of recorded numbers hybrids are much less common than in the vascular plants, but probably most have been overlooked. In the bryophytes they are usually sterile, so they exist only as hybrid sporophytes upon a female gametophyte and are relatively difficult to detect. The fact that they have been recorded in a quite wide range of bryophytes, and also of algae, suggests that hybridization in an important phenomenon in all groups of plants (as well as in fungi).

Intergeneric hybrids are much less common. In the British vascular plant flora there are only 34 interspecific combinations of intergeneric hybrids known, involving only 17 intergeneric combinations. Extrapolated to a world scale, this indicates that there might be about 2 930 intergeneric hybrid combinations. But again, there are many more artificially produced intergeneric hybrids, and natural intergeneric hybrids occur in algae and bryophytes as well as in the vascular plants. 'Wide' hybridization·is particularly a feature of the *Orchidaceae*, where artificial hybrids involving as many as eight different genera in one offspring have been synthesized. In the Poaceae, another family in which intergeneric hybrids are rather frequent (although here, unlike the Orchidaceae, they are mostly sterile), all naturally occurring hybrids are between genera within one tribe, although intertribal crosses have been artificially synthesized. On the other hand no hybrids have so far been discovered or synthesized between plant species assigned to different families.

Little generalization can be made concerning the frequency of hybridization, since between extreme cases there is every intermediate. At one

extreme one may cite the case of the grass *Catapodium marinum* ×
C. rigidum, known only as a single plant found in 1960 in Wales.[30] Both
parents and the hybrid are annuals, so that this individual lived for only a few
months. At the other extreme come hybrids such as *Geum rivale* ×
G. urbanum, *Betula pendula* × *B. pubescens* and *Circaea alpina* ×
C. lutetiana, which appear to occur whenever the two parents come into
contact, and which in many regions have become more common than one or
even both parents. Naturally, the majority of hybrid combinations come
somewhere between these extremes, and most nearer the former situation,
but there are certainly a great many very common and widespread hybrids.

The prospects of artificial hybridization at one time seemed vastly greater
when the technique of *somatic hybridization* was introduced. This involves the
fusing of single cells from two species in cell-culture, and the raising of a
hybrid plant from this fusion product. If the two cells were diploid the
resultant hybrid is already an amphidiploid, but if haploid cells (e.g. from
anther-culture) are used a diploid hybrid results. Despite the early promise of
the technique, since the production of the first mature somatic hybrid in 1972
only 27 further such hybrids have been obtained. Moreover 26 of these 28
successful experiments involve Solanaceae, and 18 of them are within the
genus *Nicotiana*.[121] The other two somatic hybrids are within *Daucus*, and to
date no intergeneric combinations have been grown to maturity. However,
wider crosses might be attainable with improved techniques.

The recognition of hybrids

One might expect a hybrid to be intermediate in appearance between its
parents, and usually this is broadly true. However, it is not always so, and in
any case intermediacy occurs in many degrees and guises. A number of
criteria can be applied to assess the likelihood that a plant is a hybrid:

1   *Phenetic intermediacy between the putative parents*
Usually intermediacy is sought in terms of morphological characters, since
these are most easily recorded; both quantitative and qualitative characters
are of value. A wide range of biometrical techniques is now available by
which one can quantify the characteristics of a series of plants. If the scale is
chosen carefully the two putative parents can be so scored that they fall far
apart on the scale, and therefore intermediates fall somewhere between
them. The simplest of these techniques are the *hybrid index* and the
pictorialized scatter diagram introduced by Anderson[9] in the 1930s and
1940s—the former as a histogram (Fig. 6.2A) and the latter in the form of a
two-dimensional scatter diagram (Fig. 6.2B). The hybrid index is obtained by
selecting a number of characters by which the two species differ, assigning the
score 0 to each of the attributes of one parent, and 2 to each of the opposite
attributes of the other parent, and summing the scores for each plant. The
hybrid index of plants typical of one parent will thus be 0, and of those of the
other parent 2N, where N is the number of characters. Intermediates will of
course fall somewhere between the extremes. If the characters used are
quantitative, the scale can be extended from 0–2 to, say, 0–6. In the
pictorialized scatter diagram each of the points is embellished with symbols

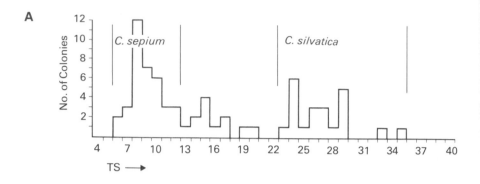

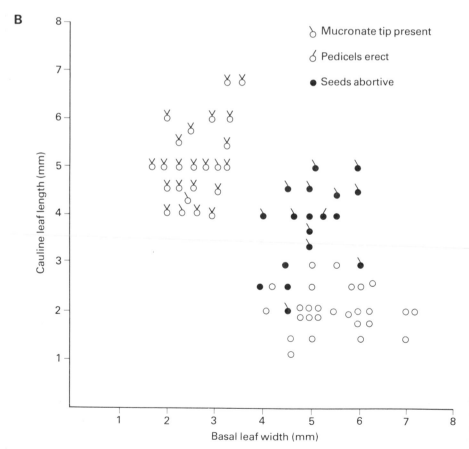

Fig. 6.2 A, Histogram of mean hybrid indices (TS) of 70 different populations of *Calystegia*, involving *C. sepium* (left), *C. silvatica* (right) and their hybrids, redrawn from Stace.[401] **B,** Pictorialized scatter diagram of herbarium specimens of *Luzula*, involving *L. forsteri* (top left), *L. pilosa* (bottom right) and their hybrids, redrawn from Ebinger.[107]

designating diagnostic attributes. The two axes are suitable quantitative characters chosen from the list of diagnostic characters separating the two parental species.

These simple techniques have proved extremely effective in a great many instances, and there are numerous examples in the literature demonstrating their efficiency in indicating the morphological intermediacy of individuals or taxa. Nevertheless, they are obviously extremely subjective, and many alternative but more complicated methods have been devised with a view to eliminating this subjectivity. Most of these involve the weighting of each character to a degree commensurate with the extent of its discrimination between the taxa, i.e. 'good' taxonomic characters are given more weight than 'bad' ones, and of necessity involve a computer. Examples are the use of discriminant analysis to separate species and hybrids of *Gentianella*,[331] the use of canonical analysis to construct a 'weighted hybrid index',[186] and the combination of principal components analysis and cluster analysis to evaluate hybridization in *Quercus*.[368]

By no means always, however, do more sophisticated methods produce better results.[159] Moreover, the demonstration or not of intermediacy does not necessarily indicate whether one is dealing with a hybridizing situation. If two species differ by fairly simple qualitative characters, the hybrid between them will inherit each of the characters from one parent or the other according to the dictates of Mendelism. The net result is likely to be that the hybrid will possess some characters of one parent, some of the other, and some intermediate. By chance this will sometimes result in a hybrid which is not halfway between its two parents, but much nearer to one than the other, or even indistinguishable from one, for example *Geranium purpureum* × *G. robertianum*, which is morphologically identical with the latter species. In some plants there are genetic mechanisms which ensure that hybrids are much closer in appearance to the female parent, a phenomenon known as **matrocliny**, for example many roses. Many cases are also known where a hybrid possesses characteristics not found in either parent, or lacks a characteristic common to both parents, for example yellow petals in a hybrid between two white-petalled species of *Ranunculus*.

These principles and problems can be demonstrated equally with regard to anatomical (Fig. 6.3), chemical and chromosomal characters, which are often used to identify hybrids. Aspects of the use of chemical compounds in detecting hybrids are mentioned in Chapter 4, where the additive nature of their inheritance was shown to be of value.

In hybrids between species with different chromosome numbers an intermediate number is to be expected, and usually is found. Hence *Brassica oleracea* ($2n = 18$) × *B. rapa* ($2n = 20$) has $2n = 19$, and *Dactylorhiza fuchsii* ($2n = 40$) × *D. maculata* ($2n = 80$) has $2n = 60$. Owing to the occasional formation of unreduced gametes by some species and to their chance participation in fertilization, hybrids can be produced which do not have intermediate chromosome numbers. *Betula pendula* ($2n = 28$) × *B. pubescens* ($2n = 56$) has $2n = 42$ or 56; *Festuca pratensis* ($2n = 14$) × *Lolium perenne* ($2n = 14$) has $2n = 14$ or 21.

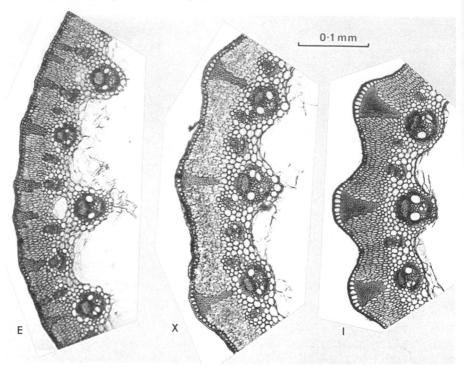

Fig. 6.3 Photographs of stem sections: **E**, *Juncus effusus*; **I**, *J. inflexus*; **X**, *J. effusus* × *J. inflexus*, showing the intermediacy of the hybrid. Photographs by A. N. Scott.

2   Reduced fertility

Interspecific hybrids range from being absolutely sterile to being as fertile as either parental species. The water buttercup *Ranunculus hederaceus* × *R. omiophyllus*, for example, is completely sterile, the stamens and carpels aborting before meiosis would have occurred.[71] *Silene maritima* × *S. vulgaris* shows no reduction in fertility,[271] and indeed these two taxa are often united as two subspecies of one species. Many other hybrids show very high levels of fertility, for example *Geum rivale* × *G. urbanum*, *Salix caprea* × *S. cinerea* and *Lolium multiflorum* × *L. perenne*, though usually some reduction in fertility is measurable and in the majority of hybrids sterility is well marked. Nevertheless the number of situations which have been exhaustively investigated and in which fertility has been demonstrated to be 0% is very small. Fertility of 0.1% or even less is highly significant in evolutionary terms, and it can be reasonably argued that the great majority of hybrids which reach maturity are fertile to some degree.

In fact hybrids showing the whole range of sterility mentioned in the last paragraph represent only one part of the spectrum of hybrid success; the other part is illustrated by hybrids which die at the zygote, embryo or seedling stage, or which become mature individuals but produce no flowers. Two complicating factors should also be mentioned. Firstly, there are many causes of reduced fertility other than hybridity, for example environmental effects, or lack of a compatible pollen-source in a self-sterile species. Secondly, the normal use of pollen-stainability as an indication of fertility on the male side (and hence overcoming the latter of the two above complications) is additionally open to the objection that in many hybrids, for example most in the genus *Juncus*, apparently perfect pollen is produced but it will not germinate. The use of so-called 'vital stains' does not altogether overcome this problem. The study of chromosome behaviour at meiosis (pollen or megaspore formation) often demonstrates abnormalities which may result from unlike genomes having been brought together in a hybrid, but these can also be the result of mutations or of environmental phenomena.

3   *F₂ segregation*

It is commonly known by gardeners and plant breeders, as well as by taxonomists, that whereas an F_1 generation might be relatively constant in appearance, the next generation (F_2, formed from breeding within the F_1 generation) might be very variable, the various characters of the parents segregating out to give a wide range of variation. At its extremes this range of variation might even mimic the appearance of one or both parents, or it might show new combinations of attributes. This topic has been much studied by Anderson.[9] If large families of seed from suspected (probably F_1) hybrids are grown in an experimental plot, the appearance of divergent segregants in the resultant F_2 generation (Fig. 6.4) is reasonable evidence that the parent was in fact a hybrid, although the absence of segregants does not necessarily indicate the absence of hybridization.

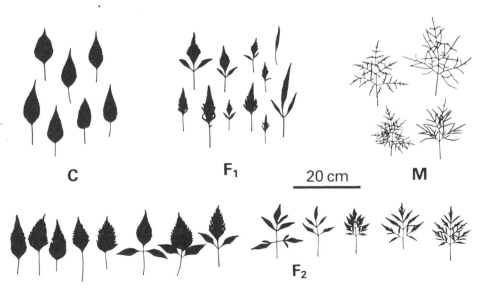

Fig. 6.4 F_2 segregation in leaf characters of the hybrid *Bidens ctenophylla × B. menziesii* from Hawaii, adapted from Mensch and Gillett.[288] Leaf-outlines of *B. ctenophylla* (**C**), *B. menziesii* (**M**), F_1 and F_2 populations.

4   Distributional evidence

In many genera which frequently form interspecific hybrids some of the most useful circumstantial evidence for the identification of hybrids comes from a study of the species present in the locality. Those found near to the hybrid are obviously more likely parents than those found at some distance. Such considerations are particularly valuable where two different hybrid combinations give extremely similar offspring, for example *Potentilla erecta* × *P. reptans* and *P. anglica* × *P. reptans*. On the other hand, hybrids often do occur well away from one or even both parents, owing to the long-distance dispersal of pollen (e.g. *Quercus petraea* × *Q. robur*), the subsequent extinction of one or both of the parents from the area (e.g. *Gentianella amarella* × *G. germanica*), or the independent migration away from the area by the hybrid (e.g. *Nasturtium microphyllum* × *N. officinale*).

5   Artificial resynthesis

If an artificial hybrid can be synthesized and is shown to resemble closely the putative hybrid in all respects, there is a strong likelihood that the latter has been correctly identified. A negative result does not, however, necessarily indicate a mis-identification. There are a great many instances where, despite numerous attempts, known wild hybrids have not been resynthesized. However many attempts may be made artificially, they must represent only a tiny fraction of the natural interspecific pollinations, and of the combinations of genotypes which have met in the wild. The chances that compatible strains have been used in attempts at resynthesis, and under the correct experimental procedure and conditions, are obviously low in some cases. Moreover, even where hybrids have been obtained, they might sometimes not resemble very closely the natural hybrid, because of the different parental genotypes involved. There are many known examples of hybrids which display large differences according to the combination of genotypes involved, or to the sexual direction of the cross. Such differences also exist in natural hybrids of a given species combination, and may provide a further obstacle to the successful recognition of hybrids.

It is thus clear that there is no invariable formula or set of clues for the identification of hybrids. Each case has to be judged on its own merits, using as many of the above five criteria as possible. Neither is it possible to extrapolate from one situation to another, even within a closely related group of taxa. It is possible to say, with *hindsight*, that all known hybrids in a particular genus are, say, sterile, but it is not possible to predict with certainty that a particular hybrid combination will be sterile just because all the others within the same genus are. The phenetic characteristics of a hybrid are equally unpredictable; a character which is dominant in one cross might be recessive in another. Thus interrupted stem-pith is dominant over continuous stem-pith in the hybrid *Juncus balticus* × *J. inflexus*, but recessive to it in the hybrid *J. effusus* × *J. inflexus*.

Isolating mechanisms

The facts that many hybrids which do not occur in the wild have been synthesized experimentally, and that hybrids known in the wild do not always occur wherever their two parents grow close together, show that there must be factors operating in the wild which prevent hybridization. It is possible to divide these into *prezygotic mechanisms* (1–9 in the list below) and *postzygotic mechanisms* (10–15) according to whether they operate before or after sexual fusion. The former may be subdivided into *pre-pollination mechanisms* (1–7) and *post-pollination mechanisms* (8–9). Alternatively mechanisms 1–7 and 15 may be described as *external* and 8–14 as *internal*. The following classification follows that of Stace,[405] who provides a detailed discussion. Here each mechanism is defined, an example given where it is known to operate, and an indication of the causes of its likely breakdown, if any, given.

1   Geographical Isolation   Two species are separated geographically, so that they do not come into contact. Example: *Larix decidua* in Europe and *L. kaempferi* in Japan. These species hybridize readily in cultivation.

2   Ecological Isolation   Two species occur in the same general area but are separated ecologically, occupying different habitats. Example: *Silene alba* on light soils in open places and *S. dioica* on heavy soils in shade or in areas of high rainfall. Their habitats overlap in some cases, especially on roadside verges, and hybridization is then common.

3   Seasonal Isolation   Two species occur in the same localities but flower at different seasons. Example: *Sambucus racemosa* and *S. nigra*, the latter flowering 7–8 weeks later than the former. Hybrids occur in exceptional seasons when the flowering periods overlap.

4   Temporal Isolation   Two species flower during the same period, but the pollen is released and/or the stigmas are receptive at different times of the day. Example: *Agrostis tenuis* and *A. stolonifera*, the former flowering in the afternoon and the latter in the morning. Pollen release at an unusual time of day due to abnormal weather conditions, or pollen remaining on insects or airborne for long periods, leads to hybridization.

5   Ethological Isolation   Related, interfertile species are pollinated by different sorts of pollen vectors, or on any particular forage an individual animal keeps to one species only. Example: *Aquilegia formosa*, pollinated by humming-birds, and *A. pubescens*, pollinated by hawkmoths. Hybrids arise from the inconstancy of vectors, especially when flowers of their preferred species are rare.

6   Mechanical Isolation   Related, interfertile species have differently structured flowers which makes it difficult for pollen vectors to transfer pollen to the stigmas of flowers other than those of the same species from which it was obtained. Example: *Ophrys insectifera* and *O. apifera*. Hybridization can result from the attentions of 'clumsy' insects, often those which do not usually visit either of the species concerned.

7   Breeding Behavioural Isolation   Species which are predominantly inbreeding or apomictic hybridize far less frequently than species which are self-incompatible. Example: *Festuca rubra* × *Vulpia fasciculata*, the latter parent being semi-cleistogamous and largely self-fertilizing. Hybrids with this species as the female parent are easily made artificially.

8   Gametophytic Isolation   Cross pollination takes place, but the pollen tube fails to germinate or to reach and penetrate the embryo sac of the female parent. This is *by far* the commonest isolating mechanism. If the stigmas of almost any plant are examined microscopically pollen grains from other species will be found adhering to it, yet not germinating. Many of these species are of course only remotely related to the recipient plant. When the relationship is closer the pollen may germinate but fail to penetrate the embryo sac. Examples are found in cereals, *Lathyrus* and *Datura*. The mechanism can be overcome in some cases by stylar surgery.

9   Gametic Isolation   The pollen tube releases the male gametes into the embryo sac, but gametic fusion and/or endospermic fusion does not take place. This has been studied in various crop plants and other genetically well-known genera.

10   Seed Incompatibility   The zygote or immature embryo ceases development, so that a mature seed is not formed. Example: *Primula elatior* × *P. veris*. This phenomenon is very well known to plant breeders. It is often, though not always, associated with failure of the hybrid endosperm to provide adequate nourishment for the developing embryo, in which case it can often be overcome by excising the embryo and culturing it on nutrient media.

11   Hybrid Inviability   Germination occurs, but the resultant F_1 hybrid dies some time before its production of flowers. Example: *Papaver dubium* × *P. rhoeas*. This can also sometimes be overcome by culture techniques.

12   Non-fitness of F_1 Hybrids   The F_1 hybrids reach maturity (or have the potential for doing so), but are not successful competitors in the wild. Usually this relates to the availability of suitable ecological niches, which may be different from those of either parent, and it is often coupled with ecological isolation, as in *Geum rivale* × *G. urbanum*.

13 F_1 Hybrid Sterility   The F_1 hybrids are fully viable but sterile, and therefore do not contribute to future generations. Sterility may be manifested from early (e.g. flower-bud abortion) to late (e.g. F_2 embryo abortion) stages. Example: *Senecio squalidus* × *S. vulgaris*. This isolating mechanism is often, though by no means always, associated with poor chromosomal pairing at meiosis, especially when the parents have different chromosome numbers.

14 F_2 Hybrid Inviability or Sterility   In these situations the disability is delayed until the F_2 generation, or even later. The F_1 hybrid *Festuca rubra* × *Vulpia fasciculata* is largely (over 99%) sterile, but a few plants have been obtained from it. These are weak and do not flower profusely.

15   Non-fitness of F_2 Hybrids   This is again the case of disability delayed until a later generation. This isolating mechanism is exemplified by those many situations in which two species are fully interfertile and produce fully fertile hybrids, yet the two parents remain recognizable as distinct entities and are often both more common than the hybrids. For various reasons the backcross, F_2 and later generations are less successful in nature than the parental species. After a few generations a multitude of hybrid offspring might have been produced, demanding a multitude of ecological niches. The likelihood that all these requirements will be met is not high.

Although these mechanisms have been catalogued separately for convenience, it is important to realize that often they do not occur in isolation. In many, probably most, cases several mechanisms operate to keep any particular pair of species effectively isolated. Especially effective are mechanisms which operate at different levels, for example pre-pollination, post-pollination and postzygotic. The much-cited example of *Geum rivale* and *G. urbanum* could be used as examples for at least six mechanisms (numbers 1, 2, 3, 5, 12 and 15).

The occurrence of hybridization is often referred to as the breakdown of isolating mechanisms. Such a phenomenon, the hybridization of two once-distinct species, was referred to by Mayr[275] as *secondary intergradation*. In other instances hybridization is a manifestation of *primary intergradation*, genetic exchange between two entities which have not yet evolved perfect isolating mechanisms. The distinction between these is of evolutionary significance, but doubtfully of great taxonomic importance.

Consequences of hybridization

The existence of hybrids between two species can cause practical taxonomic problems, because such plants are not identifiable with any one species. If they remain infrequent and occur only where the two parental species meet they are nearly always readily recognizable as hybrids and the problems which they cause are minimal, especially when they are highly sterile. But in many cases they are fertile, and produce F_2 and backcross generations, which often become common and widespread. Indeed, they can become commoner than either partent, for example various combinations in *Quercus*, *Betula*, *Ulmus*, *Rosa* and *Crataegus*.

Fertile hybrids may give rise to *hybrid swarms*, where by backcrossing and the production of F_2 and later generations the parental species become connected phenetically by every possible intermediate type, so that one species grades almost imperceptibly into the other (Fig. 6.5A). The parental limits can sometimes be detected by studying their variation in areas where only one of them occurs, or where they both occur but do not hybridize.

The existence of hybrid swarms indicates that there is a spectrum of ecological niches available to satisfy the requirements of a wide range of hybrid offspring. It is commonly observed that hybrid swarms (in fact hybrids in general) are particularly common in recently disturbed habitats, in which many new micro-niches have been artificially but unwittingly established. This was termed by Anderson[8] 'hybridization of the habitat'. Often, however, this is not so, and hybrid swarms are not able to develop: *introgression* (or *introgressive hybridization*) may take place in such situations. This term was introduced by Anderson and Hubricht[10] to describe the repeated backcrossing of a hybrid to one or the other parent, the hybrid products coming to resemble that parent quite closely after even a few generations but differing from it by some characteristics of the other species (Fig. 6.5B). Anderson and

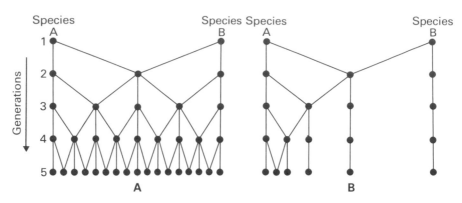

Fig. 6.5 Diagrammatic representation of: **A**, hybrid swarm formation and **B**, introgression. As in Fig. 6.1, the dots represent individuals whose ancestry is indicated by the lines.

Hubricht spoke of 'an infiltration of the germplasm of one species into that of another'. On the grounds that the ecological requirements of a hybrid phenetically closely resembling one parent are most likely also to resemble the requirements of that parent, introgression becomes a more likely outcome than the formation of hybrid swarms in situations where there are habitats suitable for the parents but not for intermediates.

Within a limited area introgression usually takes place only from one species to the other, but there are many examples known (e.g. *Quercus petraea × Q. robur*) where introgression occurs in one direction in some areas and in the opposite direction in others. Moreover, the same pair of species can in other situations give rise to hybrid swarms. It seems to be largely the effect of environmental factors which decide whether hybrid swarms or introgression (and the direction of the latter) result. These factors are not simply the presence of different ranges of ecological niches, but include the habitat or pollen-source preferences of pollen vectors, and the sexual direction of the original cross (hybrids are likely to occur close to, and therefore more likely to backcross with, their female parent).

Introgression often occurs when the F_1 hybrid is relatively infertile, presumably because it is then much more likely to backcross to a fully fertile parent than to mate with a second infertile F_1 hybrid to produce an F_2 hybrid. F_1 hybrids which are rare are also more likely to backcross than to produce an F_2 generation. Indeed, F_1 hybrids which are rare, ill-adapted to their environment and largely sterile appear to be very effective bridges for interspecific gene-flow. Moreover, the products of hybridization are likely to be most successful in evolutionary terms when they closely resemble one or other parent, as they then simply act as a means of enriching the gene-pool of an already successful species rather than producing a totally new sort of individual.[9]

In some areas hybridization between two species has become so prevalent that the majority of individuals are of hybrid origin, and in extreme cases 'pure' examples of one or even both species no longer occur in that locality. One often reads or hears of species being 'hybridized out of existence'.

Hybridization probably does sometimes assume such an aggressive role, but hybridization by long-range dispersal of pollen, or the extinction of one of the parents for reasons not related to hybridization, can produce the same end results.

Stabilization of hybrids

Although the occurrence of hybridization often obscures the distinction between species, in other cases it gives rise to new entities which are treated as separate species. This topic is often considered under the heading of the *stabilization* of hybrid progeny. It can arise in a number of ways, of which Grant[164] enumerated seven: (1) vegetative propagation, (2) agamospermy, (3) translocation heterozygosity, (4) unbalanced polyploidy, (5) amphidiploidy, (6) recombinational speciation, and (7) hybrid speciation. Grant discussed these seven phenomena under five headings: the clonal complex (1), the agamic complex (2), the heterogamic complex (3 and 4), the polyploid complex (5), and the homoploid complex (6 and 7).

The most well-understood of these is *amphidiploidy* (*polyploid complex*), discussed in Chapter 5. As explained, it involves the formation of a new fertile taxon, which behaves as a diploid, by the doubling of the chromosome number of a sterile interspecific hybrid. If one includes all the diploidized segmental allopolyploids mentioned in the previous chapter, it seems likely that between 30% and 70% of all vascular plant species arose in this way.

Simple F_1 hybrids have often become stabilized in various ways so that they have been regarded by taxonomists as species. Stace[405] considered this treatment advisable whenever hybrids 'have developed a distributional, morphological or genetical set of characters which is no longer strictly related to that of their parents'. In other words, if the hybrid has become an independent, recognizable, self-reproducing unit, it is *de facto* a separate species. Totally or highly sterile hybrids have often become extremely successful species by adopting either a vigorous mode of vegetative or asexual reproduction or an agamospermous mode of pseudo-sexual reproduction. These are the *clonal complex* and *agamic complex* respectively of Grant, and are discussed later in this Chapter under 'Apomixis'.

Under his heading *homoploid complex* Grant placed all those mechanisms which involved no loss of sexuality and no change of ploidy level or genetic mechanism. For example, in the genus *Gilia*, there are a number of well-defined diploid species of which some (e.g. *G. achilleaefolia*) seem undoubtedly to be of hybrid origin. But in this genus there are a number of other 'hybrid species' which have become tetraploids (e.g. *G. clivorum*), and in most examples of homoploid complexes polyploidy seems also to play a part. In *Euphrasia*, Yeo[483,484] recognized three consequences of hybridization: hybrid swarms, incipient speciation and introgression, of which the second is relevant here. Incipient speciation in this genus involves the formation of 'extensive populations of comparatively uniform plants apparently of hybrid origin', usually a diploid and a tetraploid giving rise (via a triploid which produces some haploid gametes) to a new diploid (Fig. 6.6), although some examples all at the diploid level are known. Ubsdell[444] discovered in Lancashire a new fertile allohexaploid species, *Centaurium intermedium*, which

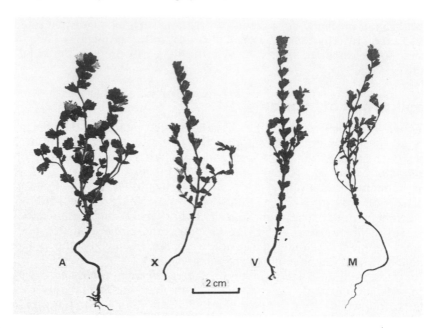

Fig. 6.6 Photographs of herbarium specimens of the diploid *Euphrasia anglica* (**A**), the tetraploid *E. micrantha* (**M**), a fertile diploid hybrid between them (**X**), and the diploid *E. vigursii* (**V**), thought to be a locally successful variant of the last.

she believed arose by the backcrossing to one of its parents of a largely infertile tetraploid hybrid between two tetraploid species. Presumably the hexaploid level arose due to the formation of some unreduced (tetraploid) gametes.

A number of examples of species and lower taxa which have arisen as introgressed variants of established species have been described, mostly in North America. One of the most striking is the postulated origin of *Purschia glandulosa* from the introgression of *Cowania stansburiana* into *Purschia tridentata*, involving two separate genera of woody Rosaceae.[419,428]

Grant's *heterogamic complex* is represented by two relatively unusual mechanisms by which hybrid plants can discover a successful formula for adaptation and survival: *translocation heterozygosity* and *unbalanced polyploidy*. The best examples of the first are found in many species of *Oenothera* which form multivalent rings of chromosomes at meiosis (see p. 124), and of the second in *Rosa*, where pentaploid taxa are maintained by the female parent contributing four sets and the male parent only one set of chromosomes to the zygote.

There are good grounds for considering hybridization as one of the major evolutionary mechanisms in plants. It seems plausible that a phase of hybridization is followed by one of stabilization, and that this pattern is repeated cyclically.[110] Hybridization represents a period of maximum recombination, and stabilization the selection of the fittest recombinants so formed. Fortunately for botanists, different groups of plants occupy different parts of this cycle at any one period, so we are able to study all aspects of it at the present time.

Hybridization as specific or generic criteria

The existence of a large number of taxonomically 'ideal' species has led, rather naturally, to the notion that species might be defined more precisely by reference to their genetical isolation rather than to their morphological limits. Taxa which can interbreed (at least to form fertile hybrids) might be considered to represent a single species, while those which cannot might be recognized as separate species, with little or no regard to the degree of phenetic divergence. This is the basis of the so-called ***biological species***. 'Ideal' plant species compare closely with the great majority of species of animals, in which there is similarly a coincidence of the species defined in phenetic and genetical terms, and most animal taxonomists today still adhere at least partly to the 'biological species' definition. This is particularly attractive because it offers the possibility of using a single unambiguous characteristic to define species limits when a definition based on the usual (phenetic) characteristics is problematical.

Among botanists, however, there is now little adherence to such a naïve definition of a species. This is because in plants there are too many examples to consider them as atypical or abnormal of taxa in which the morphological and genetical limits do not coincide—where either the genetical limits are much wider than the morphological, or *vice versa*. In a great many groups of plants morphologically distinct species capable of hybridizing with others are the *usual* situation, and cannot be considered in any way abnormal. There would be no point in uniting large numbers of species in such genera as *Euphrasia*, *Dactylorhiza* or *Salix* into a huge, unwieldy, heterogeneous taxon on the sole ground that its members could freely interbreed. Moreover, there is no reason to believe that such non-ideal, i.e. taxonomically problematical, species are in any way biologically non-ideal. It is difficult to envisage any evolutionary disadvantage to a group of plants in being composed of other than discrete morphological and genetical units. Indeed, the great success exhibited by many 'taxonomically difficult' plants, especially in man-made habitats, indicates that they are in fact extremely well fitted to their environment.

The ability or not to interbreed is not an absolute character and, if it were to be used as a species criterion, many conditions would need to be stipulated. In the first place, many species *can* interbreed, as proved by experiments, but in the wild do not do so, or do so only in some localities. If hybrids do occur in nature they might be fertile or sterile to any degree (Fig. 6.7), and their fertility might also vary greatly from one area to another. Even fertile hybrids might not be very successful competitively in natural environments, and their ability to backcross and produce F_2 generations might therefore be impeded. Finally, only a tiny fraction of the world's species have been examined experimentally in any way, and it is clear that any attempt to redefine the species strictly on genetical criteria is quite impracticable.

The ability to hybridize is one measure of close relationship and, as with all characters, it becomes taxonomically more valuable or critical when it can be

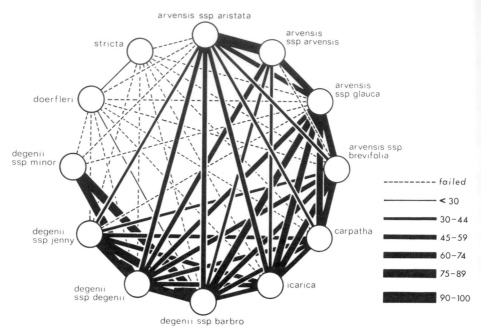

Fig. 6.7 Chart showing the percentage fertility of hybrids between various taxa of *Nigella* from the Aegean region, taken from Strid.[426]

used differentially. If hybridization in a particular genus is rare, those pairs of species which do hybridize are more closely related on at least that basis. There are then convincing grounds for expressing their relative genetic closeness in taxonomic terms, so long as they are not phenetically less similar than other pairs or groups of species in the genus. *Silene maritima* and *S. vulgaris*, and *Medicago falcata* and *M. sativa*, are two pairs which are often so treated on these criteria. This does not amount to using genetical data to over-ride phenetic evidence, but to seeking out a character, in this case genetical in nature, that provides a differential indication of relationship between the species of a genus.

Even though intersterility has been generally abandoned as a *species* criterion, it is still sometimes used as a means of delimiting genera. It could be argued, for example, that species which interbreed in the wild should be placed in the same genus, and proposals to amalgamate two or more genera (e.g. the ferns *Phyllitis*, *Ceterach* and *Asplenium*) are often made on this basis. It is to be admitted that the great majority of intergeneric hybrids involve closely related genera, such as the three above, but the *special* use of this criterion to delimit genera is totally at variance with the principles propounded in this book and it cannot be advocated. Moreover, in the Orchidaceae and Poaceae various groups, each of several very distinct genera, hybridize in the wild, and their amalgamation would result in large, heterogeneous genera of little practical value. Various suggestions that the generic limits of the grasses *Festuca*, *Lolium* and *Vulpia* should be altered to reflect the limits of natural hybridization, e.g. by amalgamating all three, or by splitting *Festuca* and amalgamating the two parts with *Lolium* and *Vulpia* respectively, not only fall foul of the above two objections but also ignore the results of increasingly extensive artifical crossings, which show that natural hybridization does not present the complete picture.[16]

Semi-cryptic species

A quite different kind of taxonomic problem from that associated with hybridization is represented by *semi-cryptic species*, so-called because their differences are marked in anatomical, chemical, cytological or genetical characters rather than morphologically. In some cases they represent incipient species, but in others they appear to be long established taxa which have not diverged markedly in gross morphology. They form the opposite end of the spectrum from many groups of Orchidaceae, in which genetically defined units can encompass an extremely wide and heterogeneous range of morphological species.

Because of their semi-cryptic nature, there is often considerable taxonomic argument concerning the correct status of the ultimate taxa; the species recognized by some taxonomists become relegated to subspecies or even varieties by others. This topic is discussed in Chapter 8. The existence of these semi-cryptic taxa appears to be based on three sorts of genetic situation, which will be discussed in turn.

1   Outbreeders with internal barriers

Many species or species complexes are known in which there are internal breeding barriers, i.e. where different races are genetically unable to interbreed, at least to form fertile hybrids. In most instances, the underlying cause is chromosomal—usually a difference in chromosome number, but sometimes in chromosome structure (*structural hybridity*). Several examples of pairs or groups of phenetically similar taxa different in chromosome number are given in Chapter 5. Others are *Empetrum nigrum* ($2n = 26$) and *E. hermaphroditum* ($2n = 52$), *Monotropa hypophegea* ($2n = 16$) and *M. hypopitys* ($2n = 48$), *Eleocharis palustris* subsp. *microcarpa* ($2n = 16$) and subsp. *vulgaris* ($2n = 38$), *Lamiastrum galeobdolon* subsp. *galeobdolon* ($2n = 18$) and subsp. *montanum* ($2n = 36$), and *Veronica triloba* ($2n = 18$), *V. sublobata* ($2n = 36$) and *V. hederifolia* ($2n = 54$) (Fig. 6.8).

In other cases, however, no obvious differences exist in chromosome number or structure, although it is difficult to be certain that there are not underlying minute structural differences (*cryptic structural hybridity*). The two taxa *Anagallis foemina* and *A. arvensis* both have $2n = 40$ and are extremely similar phenetically, but they are intersterile. In the *Cerastium arvense* group it is often more difficult to synthesize hybrids between different races, even within one ploidy level, than to produce hybrids between *C. arvense* and other species.

2   Inbreeders

Early in this chapter the idea was put forward that inbreeding species tend to generate small intra- but relatively wide inter-populational variation. Figure 6.1A represents a hypothetical extreme situation in which there is 100%

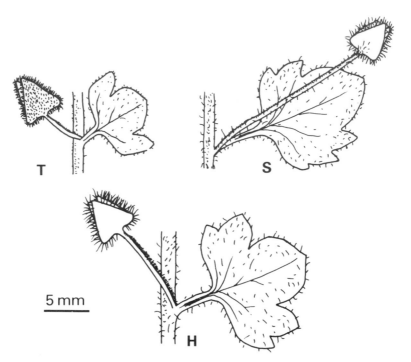

Fig. 6.8 Drawings of diagnostic parts of three semi-cryptic species of the *Veronica hederifolia* complex: the diploid *V. triloba* (**T**), the tetraploid *V. sublobata* (**S**), and the hexaploid *V. hederifolia* (**H**). Taken from Fischer.[134]

inbreeding, and where each genetically distinct entity (***biotype***) is a pure line. Whereas one can demonstrate very low levels of outbreeding (e.g. less than 0.01% in *Vulpia microstachys*), it is, of course, impossible to prove that any taxon is exclusively inbreeding. In practice the extent of inbreeding varies from 0% (or very close—even in most dioecious species very rare monoecious variants are known) to 100% (or very close). Accordingly the variation pattern varies in all degrees between those depicted in Figs 6.1A and 6.1C. Jain[226] listed seven evolutionary consequences of inbreeding, several of which bear upon the degree of intra- and inter-populational variability.

In many species the pattern lies much closer to that of Fig. 6.1A than to that of Fig. 6.1C, and the different population-types are often recognizable as distinct morphological entities. It is hardly surprising that in many genera these entities have been given taxonomic status as species or infra-specific variants. The taxa so recognized are usually capable of hybridizing, and the products of hybridization are mostly fertile and exhibit randomly recombined parental characteristics. Nevertheless, their innately low level of interbreeding is frequently reinforced by marked ecological or geographical separation, for each biotype is usually well adapted to its local habitat, and the field botanist can become quickly acquainted with them.

As examples of predominantly inbreeding species composed of local populations which have each been given specific rank, one may quote *Vulpia microstachys* in western North America (*c.* 8 taxa named), *Senecio vulgaris* (*c.* 12 taxa named in Britain), *Capsella bursa-pastoris* (*c.* 70 taxa named in north-western Europe), and *Erophila verna* (over 200 taxa named in western Europe, mainly France). These taxa of a lower level differ in small characters of the type whose inheritance has been much investigated in classical genetical studies: hairiness of various parts, degree of branching, shape of pod, degree of leaf-dissection, etc. In an outbreeding species these characters would be found in all combinations, the precise proportions of which usually change from generation to generation and habitat to habitat. All the above-named species are annuals or ephemerals of short-lived habitats and several of them are very successful weeds. They are all very largely inbreeding, and in the case of the *Vulpia* the flowers are cleistogamous (not opening to allow cross-pollination). Clearly the ability to recognize taxa of a lower order is related to the low levels of recombination which occur in these plants, and it is each species in its wider sense which is the 'equivalent' of an outbreeding species.

The taxonomic recognition of entities within inbreeding species was first seriously suggested by the Frenchman A. Jordan[241] in *Erophila verna* (Fig. 6.9), which he subjected to extensive cultivation experiments. Consequently these numerous minor taxa, for which the term *microspecies* is appropriate, became known as *Jordanons* to distinguish them from the **Linnaeons** or 'normal' species of a larger order. The Jordanons of *Erophila verna* were found by Winge[482] to represent single biotypes or groups of similar biotypes, some of which are marked by only one or two characters. Comparable work in the genus *Capsella* was carried out by Shull.[378] The situation in *Erophila* is complicated by the existence of a dysploid series with chromosome numbers

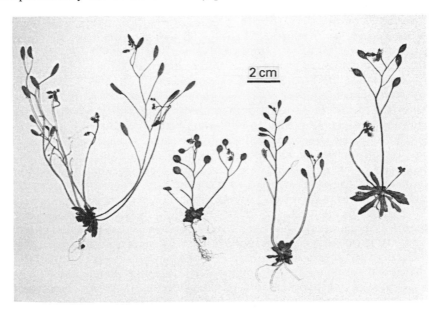

Fig. 6.9 Photographs of herbarium specimens of four microspecies of the *Erophila verna* group, each of which would have been given specific status by Jordan.

of $2n = 14$ to $2n = 94$, but this is not so in any of the other three species mentioned above.

The case against recognizing the microspecies as taxonomic species is that there are unmanageable numbers of them, and they are not usually constant over long periods. Sooner or later cross-fertilization will occur and recombinant types which themselves become new microspecies are produced. 'There is no established taxonomic principle, however, dealing with the recognition of species in inbreeding groups',[98] and the fact that today Jordanons are mostly not recognized as species is simply a reflection of the current taxonomic fashion. In fact in different genera the microspecies are recognized at different levels, from species (rarely) to forms, or not at all. In the *Erophila verna* group a compromise is often made, whereby the many microspecies are loosely aggregated into a few groups which are recognized at the species or subspecies level.

Although the taxonomy of inbreeding species has been much neglected in recent years there is every reason to believe that its placement on a firm basis would result in much benefit to those who study weedy species, as has the description of microspecies in many apomictic groups (q.v.). Population geneticists are paying detailed attention to many inbreeding species at the moment. Species which are discussed above are generally recognized as **r-strategists** (species allocating a high proportion of their resources to reproductive activity), as opposed to **K-strategists** (species allocating a high proportion of their resources to vegetative vigour and competition), but these two concepts are relative rather than absolute.[147]

In non-vascular plants the taxonomic consequences of inbreeding versus outbreeding have been scarcely studied,[382] although there are many genetical data available. A far higher proportion of non-vascular than vascular plants are obligate outbreeders, for in plants with a free-living gametophyte dioecism is common. It is present in about 80% of all liverworts and over 50% of British mosses.[358] In pteridophytes gametophytic dioecism is rare, and sporophytic dioecism unknown (as it is in bryophytes).

3   Apomictic taxa

Apomictic taxa are those which reproduce by *apomixis*—the habitual reproduction by non-sexual means. In flowering plants this can take two forms—*vegetative apomixis*, where reproduction is asexual or vegetative (N.B. only where it *replaces* sexual reproduction is it, by definition, apomictic), and *agamospermy*, where seeds are formed by pseudo-sexual means. Some authors restrict the term apomixis to mean the same as agamospermy, and in pteridophytes and lower plants, where the term agamospermy is inappropriate, apomixis is generally used to exclude vegetative apomixis. Many different mechanisms of agamospermy are known and the subject has been thoroughly reviewed by Gustafsson[171,172,173] and more briefly by Stebbins[418] and Grant.[164] In all cases the embryo is formed from entirely maternal tissue, so that the offspring are genetically identical with their female parent.

Vegetative apomixis occurs in a great variety of situations, mostly where sexual reproduction is difficult or impossible. Examples include dioecious species where only one sex is present (e.g. *Elodea canadensis* in Europe), species where propagules such as bulbils occur in the place of flowers (e.g.

proliferating or pseudo-viviparous forms of *Poa alpina* and *Allium vineale*), and species which reproduce vegetatively but are sexually sterile for genetic reasons (e.g. hybrid taxa in *Potentilla, Circaea, Mentha, Opuntia, Potamogeton, Hyacinthus* and many ferns) (Fig. 6.10). In eastern North America four fern genera exhibit vegetative apomixis of the gametophytes, which have become dispersed independently from the sporophyte and in some instances occupy far larger geographical ranges.[149] In many bryophytes sexual reproduction is rare, often because the species is dioecious and only one sex occurs in the locality. In about 14% of British mosses and a higher proportion of liverworts sporophytes have not been found in Britain, and in some taxa they are not known anywhere.[358] Vegetative apomixis is probably also very common among algae.

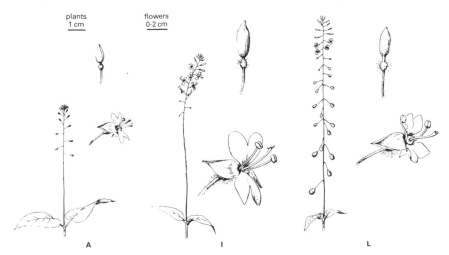

plants
1 cm

flowers
0·2 cm

A   I   L

Fig. 6.10 *Circaea alpina* (**A**), *C. lutetiana* (**L**) and *C. intermedia* (**I**, the sterile hybrid between them), adapted from Raven.[339] Although sterile, *C. intermedia* is far more widespread in Britain than is *C. alpina*, and is recognized as a species in its own right.

It is worth introducing the main terms associated with agamospermous reproduction in flowering plants. The embryo may be formed directly from sporophytic maternal tissue such as the nucellus (i.e. *adventitious embryony*), or via the formation of a diploid gametophyte by the circumvention of meiosis (i.e. *gametophytic apomixis*). The circumvention is achieved by either archesporial cells (by *diplospory*) or somatic cells (by *apospory*) developing directly into a gametophyte. This gametophyte develops by either an egg-like cell (*parthenogenesis*) or some other cell (*apogamy*) growing into an embryo.

Adventitious embryony is best known in various cultivated *Citrus* taxa. Gametophytic apomixis occurs in an extremely wide range of flowering plant families from Ranunculaceae to Poaceae, being particularly prevalent in the Rosaceae (e.g. *Rubus, Alchemilla, Sorbus*) and Asteraceae (e.g. *Taraxacum, Hieracium*). Although a male parent never contributes to the embryonic tissue, pollination is nevertheless necessary for the successful agamospermous development of the seed in some apomictic plants, for example *Rubus fruticosus* and *Ranunculus auricomus*. This phenomenon, known as **pseudogamy**, has various bases, but it is most often associated with the need for a male nucleus to fuse with a female nucleus in order to produce a functional endosperm which nourishes the developing embryo. In practical biosystematic terms it greatly complicates the proof of sexuality as opposed to apomixis by the demonstration that emasculation and the exclusion of pollen prevent seed formation; lack of seed production clearly does *not* prove sexuality.

Apogamy occurs in about 10% of ferns which have been adequately studied.[459] In these cases diploid (2*n*) gametophytes are formed, and these give rise directly to new sporophytes of the same ploidy level without sexual fusion occurring. The diploid gametophytes are usually produced by normal meiosis which is preceded by a chromosome division without segregation of the chromosomes into separate nuclei, so that the **restitution nucleus** so-formed exists temporarily at the tetraploid (4*n*) level. Apogamy has been especially well studied in the genera *Pteris* and *Dryopteris*.

Apospory (the development of vegetative sporophytic tissue into a diploid gametophyte) can be easily induced in both pteridophytes and bryophytes by wounding. In the moss *Phascum cuspidatum* it may be accompanied by apogamy, as in the ferns above, but little is known of the natural occurrence of non-vegetative apomixis in bryophytes. Parthenogenesis and other forms of apomixis are well known in various algae.

The different mechanisms of apomixis outlined above give rise to a common end-result—the preservation and propagation of pure lines by essentially vegetative propagation, which produces a pattern of variation like that illustrated in Fig. 6.1A for obligate inbreeders. In this way, therefore, apomixis could be looked upon as a most extreme form of inbreeding, since its taxonomic consequences are as might be predicted from such a breeding system. However, there is a major difference, in that pure lines of inbreeders are usually quite highly homozygous, whereas lines of apomicts are often very highly heterozygous, especially because in many cases they are hybrid in origin. Apomicts are usually composed of 'populations' (better called 'agamodemes') of extremely uniform individuals. In some cases these may represent a single genotype, but this is certainly not always and perhaps only rarely so (see below).

Differences between these agamodemes are mostly small, but they are usually quite constant and with practice a taxonomist can easily recognize them individually. Needless to say, the different sorts of agamodemes of almost all apomictic species have been given names by taxonomists, but it is a matter of debate as to the level at which these taxa should be recognized. Because they are far more stable than the microspecies or Jordanons of inbreeding taxa, the temptation to recognize apomictic microspecies (**agamospecies**) as taxonomic species is far greater. In fact for the majority of apomictic vascular plants it is the fashion to name each of the agamospecies at the specific level, contrary to the situation regarding Jordanons. In the British Isles alone about 300 agamospecies of blackberry (the Linnaean species *Rubus fruticosus*), about 250 hawkweeds (*Hieracium murorum*), about 250 dandelions (*Taraxacum officinale*), 20 whitebeams (*Sorbus aria*) (Fig. 6.11), 13 lady's-mantles (*Alchemilla vulgaris*) and 9 sea-lavenders (*Limonium binervosum*), besides others, are currently recognized. More species have

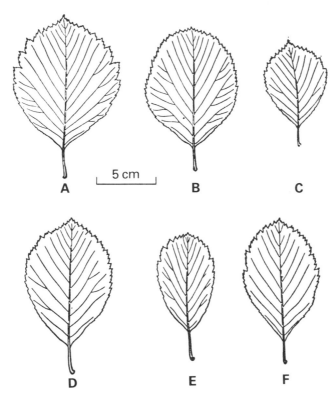

Fig. 6.11 Leaves of six representative microspecies of the *Sorbus aria* aggregate, taken from Warburg.[465] **A**, *S. eminens*; **B**, *S. hibernica*; **C**, *S. porrigentiformis*; **D**, *S. lancastriensis*; **E**, *S. rupicola*; **F**, *S. vexans*.

been described in the genus *Hieracium* (about 10 000) than in any other genus in the world. In mosses the nine microspecies in the *Bryum erythrocarpum* group[80] have probably arisen as the result of vegetative apomixis. The group is dioecious but rarely produces sporophytes; it reproduces mainly by means of gemmae produced on the rhizoids. In some apomictic groups, however, the agamospecies have usually been given infraspecific status, for example as subspecies within the *Ranunculus auricomus* group, but this reflects no more than the preferences of the specialists concerned. The use of special categories (e.g. *forma apomictica*, or *f.ap.*) has often been advocated, but scarcely ever used by practising taxonomists.

The recognition and naming of agamospecies can be justified on the grounds that the taxa so delimited differ as constantly as many sexual species, and they often have equally distinctive geographical distributions and ecological preferences. In the sense that plant names are headings under which information concerning plants can be stored, the recognition of microspecies provides a finer and therefore more useful cataloguing of information.

The distribution patterns of agamospecies vary almost as much as those of sexual species, some being very widespread and others very restricted. This is partly based upon the dispersal mechanisms of the plants concerned. Agamospecies of *Taraxacum* and *Hieracium*, which have well-adapted wind-dispersed propagules, are in general far more widely distributed than agamospecies of *Ranunculus auricomus* and *Limonium binervosum*, which have poorly developed dispersal mechanisms. Moreover the first pair are weedy genera colonizing open ground, whereas the second pair have more demanding ecological requirements and their habitats are themselves relatively uncommon and disjunct. Most agamospecies of these two genera are known from only one locality, presumably close to where they arose, yet genetically they may be of the same status as the widespread agamospecies of *Taraxacum* and *Hieracium*.

More recently evolved agamospecies are also more likely to have very restricted distributions. In *Rubus*, Newton[306] described eight grades of distribution; taxa within the more restricted grades (in extreme cases a single bush) are in general more sterile and probably of more recent origin. Newton suggested giving binomials only to plants in the first five grades and to some of category 6 (defined as containing 'taxa which may or may not turn out to be deserving of subsequent description').

The genus *Rubus* also represents in extreme form a factor which greatly complicates the issue of recognizing agamospecies as taxonomic species. Whereas many taxa in which apomixis occurs are always non-sexual (*obligate apomicts*), some may be apomictic or sexual (*facultative apomicts*). In the genus *Rubus* some taxa are facultatively apomictic, producing good pollen and a mixture of apomictic and sexual carpels, whereas a few are wholly sexual and many wholly apomictic.[185] In *Taraxacum* different plants of a single microspecies may be either wholly apomictic or wholly sexual (not both on one plant), but most of the microspecies are wholly apomictic and a few are wholly sexual.[356] In ferns, too, apogamous species usually possess functional antheridia on their gametophytes and can act as male parents in hybridizations. All apogamous ferns are, however, obligate apomicts on the female side and their gametophytes bear archegonia which are all non-functional.

Facultative apomixis causes great taxonomic difficulties, because it gives rise to plants which are variously intermediate between the erstwhile distinct but closely similar agamospecies. In genera such as *Rubus* the problem is intractable, and a rule-of-thumb as suggested by Newton is the only practical solution. The plants in Newton's groups 7 and 8 are recent hybrids, some of which will die out and others of which will become, in time, successful agamospecies in their own right.

The situation seen today in the north-west European *Rubus* flora is highly dynamic and relatively unusual. It may well be, however, a stage which all apomictic groups undergo at some time in their evolution, and perhaps a stage which is undergone not once but successively, in many cases in a cyclic manner similar to that described previously as a hybridization–stabilization cycle. The two ideas are, in fact, very closely related, since the majority of (perhaps all) agamospecies are hybrid in origin. Apomixis represents, for hybrids, an escape from sterility, which in sexual terms is, in many cases, inevitable. All the agamospecies in genera such as *Ranunculus*, *Rubus*, *Hieracium* and *Taraxacum* are polyploids, often with odd levels of ploidy (triploids, pentaploids, etc.) and usually with very irregular meiosis. In

Limonium binervosum most of the agamospecies are tetraploids lacking one chromosome ($2n = 4x - 1 = 35$). Apogamous ferns in *Dryopteris* and other genera are also hybrids which would otherwise be sterile, for example *D. pseudomas*, a triploid with $2n = 123$.

It has been suggested that unlike the hybridization cycle of sexual plants the apomictic cycle might cease to operate when the taxa become obligately apomictic and when all related sexual species die out. In those cases apomixis can be considered a reprieve rather than an escape from sterility. However, there is some debate as to whether or not this is the usual situation. The old idea that agamospecies are single genotypes, each the product of a separate hybridization followed by the onset of apomixis, has been abandoned. In *Alchemilla* it has long been known that there is much variation within many agamospecies.[461] In *Limonium*, the *L. binervosum* group is entirely obligately apomictic, yet the pattern of variation is markedly hierarchical, just as would be expected from a sexual breeding system.[224] This pattern, which must have arisen in Post-glacial times, is interpretable only by a small number (perhaps only two) of hybridizations followed by the onset of apomixis and further differentiation, presumably via mutation alone. Evidence from this and other groups strongly suggests that obligate agamospermy is by no means 'the end of the evolutionary line' as was formerly thought. The same may well be true of vegetative apomixis.[445]

7

Information from plant geography and ecology

Many aspects of the ecology and geography of plants have been covered incidentally in previous chapters. They are clearly of much relevance to plant taxonomy because each taxon exhibits a certain pattern of distribution which is one aspect of its definition. The coincidence or not of areas occupied by related taxa has a bearing on the classification of the group, especially when its evolution is taken into consideration.[296] Moreover, geographical patterns are of great practical significance because plants are collected from distinct areas, and are catalogued in herbaria and described in Floras which are based on these same areas. For logistic reasons taxonomists are as often experts on regions as on taxa.

The salient points discussed in this chapter are but a small fraction of the distributional data and concepts which form the subjects phytosociology, phytogeography and plant genecology. They are intended to show that geography and ecology are more significant to taxonomy than the inexperienced field botanist might believe when he is confronted with an identification key which asks him to differentiate between 'plant alpine' and 'plant lowland', or by an alien grass in Europe whose origin might be Australia, Africa or America.

Patterns of geographical distribution

Geographical differentiation exists between taxa at all levels and in all degrees of spatial separation. Indeed, *effective* spatial separation varies greatly in absolute terms because different taxa may possess quite different powers of migration, either as pollen or as seeds. Taxa which occupy mutually exclusive geographical areas are termed **allopatric**, and those occupying similar or overlapping areas are termed **sympatric** or partially sympatric. Closely related sympatric taxa usually show different types of genetic, ecological and structural differentiation from those shown by closely related allopatric taxa,[111] because allopatry is itself an important isolating mechanism whereas sympatric taxa can rarely rely on spatial separation.

The patterns of distribution shown by plants form part of the subject of **phytogeography**. At the species level, *Pteridium aquilinum* (the common bracken), a number of weeds of agricultural land (such as *Poa annua*) and certain aquatic species (such as *Spirodela polyrhiza*) are examples of virtually cosmopolitan plants. At the opposite extreme many species are known only from a single population, for example *Rhynchosinapis wrightii* from the Isle of Lundy in the Bristol Channel. In the case of weeds, and perhaps also of water plants, the wider-than-usual area of distribution may be explained by relatively recent but particularly effective methods of dispersal. There are many examples of algae and bryophytes with very wide distributions. In some groups, for example freshwater algae, these are also species which are very efficiently distributed, but in other cases, for example mosses which rarely reproduce sexually, a very slow rate of evolution into distinct taxa in separate areas may be indicated.

At the genus and family level (except for examples containing only one species), distribution patterns are obviously wider, but nevertheless for taxa of a given size there is a great variation in the area occupied. Many sizeable families are confined to the Americas or to Australia, for example. The genus *Eucalyptus*, with about 500 species, occurs only in Australia (apart from two or three species in adjacent Malaysia).

If the distribution of a large number of taxa is analysed (say all the species within a given area, such as the British Isles or Australia), certain geographical patterns are found to recur consistently. These patterns and the taxa which exhibit them are known as *floristic elements* and their demonstration may be a valuable method of floristic analysis. The flora of the British Isles has been analysed in this way by (among others) Matthews,[274] who recognized 16 elements, and Birks and Deacon,[36] who recognized 17. A knowledge of the floristic elements of a recently glaciated region is helpful in understanding the pathways of Post-glacial migration into it and may also aid in taxonomic decisions in cases of doubt. Perhaps the most fruitful analyses of this type, involving the deduction of migration pathways of the flora of Scandinavia, have been undertaken by Hultén,[213] who recognized 48 floristic elements and later extended his studies to the whole North Atlantic zone. More recently Raven and Axelrod[347] have provided a floristic survey of the Californian flora.

The distinction between geographical distribution (discussed above) and ecological distribution is by no means always clear. For example, taxa confined to coastal or alpine areas, and thus separated from inland or lowland taxa, can be equally considered in either category. This is of more than theoretical significance, because traditionally ecological and geographical races have often been afforded different taxonomic treatments. The small-scale variation in populations on different islands within an archipelago poses similar problems, since there is often a complete range from inter-populational variation to large-scale variation.

The choice of the visual means of illustrating plant distributions is very important.[98,414] There is now, apart from its use by vicariance biogeographers (q.v.), a trend away from the enclosure of areas of distribution on a map by a single line (Fig. 7.1A), since, at the periphery of the area, distribution is often sparse and there is no absolute cut-off line. The use of dot-maps to some extent overcomes this problem. The area may be divided into regularly spaced squares or rectangles and the presence of the taxon in each square indicated on the map by a solid spot (Fig. 7.1B). This method has been widely used in north-western Europe[323] and is now being extended to the whole of Europe.[227] Alternatively, dots may be placed on actual localities whence the plant has been recorded (Fig. 7.1C), but the advantage of the grid-square method is that it places some obligation on the part of the recorder to search every square to ascertain the presence or absence of the taxon, and it thus

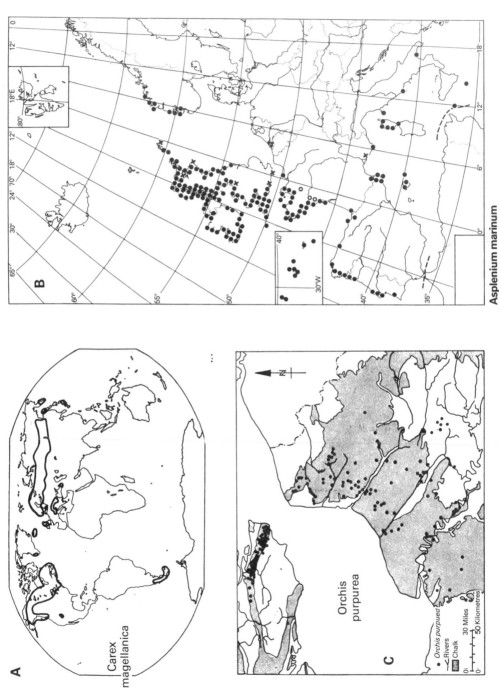

Fig. 7.1 Three styles of distribution maps: A, outline (from Moore and Chater[297]; B, dots in grid squares (from Jalas and Suominen[227]); C,

Asplenium marinum

A — Carex magellanica

B

C — Orchis purpurea

• Orchis purpued
Rivers
Chalk

0 30 Miles
0 50 Kilometres

N

ensures a reasonable level of evenness of effort over the whole area. Extra data can often be added usefully to the dots by various simple pictorial methods (Fig. 7.2).

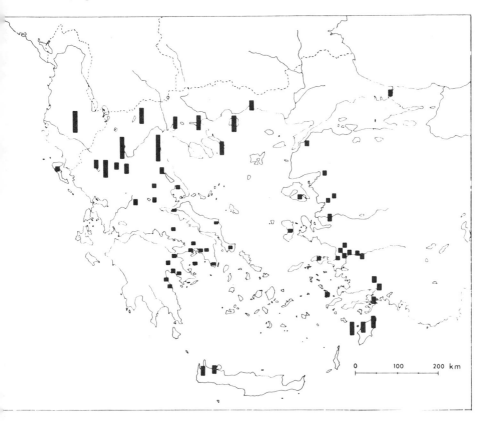

Fig. 7.2 Map of the Aegean region showing the mean length of the uppermost internode of various populations of *Nigella arvensis*, taken from Strid.[426] The height of the rectangles is relative to the internode length.

Disjunction and vicariance

Although most taxa are found fairly continuously throughout their area of distribution, some have distribution patterns which are interrupted by considerable areas from which the taxon is absent. Such taxa are said to exhibit ***disjunct*** distribution, which might have originated in several different ways. Often the disjunction has clearly arisen by the long-range dispersal of the taxon from one area to another. The presence of plants such as the coconut on a great number of far-distant volcanic Pacific islands is an obvious example. In other cases disjunctions may represent the relics of former wide, continuous distribution patterns, the intervening areas having been depopulated. It is often very difficult to distinguish between these two extreme disjunctive agents, and there are still many disputed cases. The occurrence of various alpine species on relatively isolated mountain ranges in Europe (e.g.

Alps, Pyrenees, Sierra Nevada), and the so-called Lusitanian species of south-western Ireland (which occur elsewhere only in Portugal and north-western Spain, and sometimes in Brittany and south-western England), are good examples. What appears to have been long-range dispersal could have been effected via continuous or 'stepping-stone' bridges of suitable habitats that once (but no longer) linked the two areas. The former connection of Britain to France up to only about 7 500 years ago is a classic example. Migration of species between certain European mountain ranges (e.g. Alps and Pyrenees), and similarly between various African mountain ranges, is thought to have been facilitated in past times when the climate was much colder and the present montane species were not confined to the high mountains.[299] Alternatively, plants in two separate areas might have arrived there from a third area from which they are now absent.

A further possibility is that the taxon has arisen independently in its separate areas of distribution, thus exhibiting evolutionary parallelism or convergence. Although this is usually discounted, at least when no obvious common ancestor is known, there are some instances where it is definitely known to have taken place. For example, certain amphidiploids (e.g. *Tragopogon mirus*, *Spartina anglica*) have arisen polytopically.[405]

It should also be mentioned that, when considering disjunct distributions, great care has to be taken that the disjunction is not simply the result of a plant having been carried by man to areas where it is not native. The native distribution of many weeds and crop plants is obscure because the plants have been much distributed by man, intentionally or accidentally, for several centuries, and the extent to which their distributions are naturally disjunct is unknown.

Some disjunctions have been much studied because they are spectacular or otherwise remarkable. Among these are the amphi-atlantic species found in Ireland and eastern North America (e.g. *Sisyrinchium bermudiana*),[198] and the so-called bipolar species (Fig. 7.1A) found in arctic and subarctic North America (and often arctic Eurasia as well) and in the southern tip of South America (e.g. *Carex magellanica*).[295,340] Both these cases include some examples which may not show true disjunctions, but represent the result of introductions by man. In the former case *Juncus tenuis* and *Hypericum canadense* probably represent introductions from North America, although there are botanists who would argue otherwise. The recent spread in western Ireland of *Juncus planifolius*,[372] whose nearest native area is in southern Chile, indicates how swiftly an alien species may spread even when its mode of introduction into and subsequent dispersal within Ireland is unknown. Other much studied disjunct distributions include the eastern Asia–eastern North America[177] and the South America–Australasia–South Africa[295,346] patterns. All these examples illustrate an important feature of many disjunctions, in that they are exemplified by a number of different taxa. This recurrence in a range of quite unrelated taxa of a particular pattern of disjunction reinforces the belief that this is a natural phenomenon which has to be explained in phytogeographical and biosystematic terms, invoking theories of continental drift, island formation, land bridges, adaptive radiation, and rates of colonization and evolution, etc.

A very precise taxonomic delimitation of the taxa involved in disjunctions is important. Often, on close study, the two separated elements of a disjunct taxon have been found to differ slightly, for example *Empetrum rubrum* in North and South America, *Sphaerocarpos texanus* in Europe and America, which might lead to their recognition as separate taxa. At a higher taxonomic level, the wild African bananas are now separated from *Musa* into the genus *Ensete*, and the New World and the Old World representatives of the Gesneriaceae are now largely placed in separate subfamilies. Although such discoveries might not alter the fact that the disjunction has still to be explained, they might well shift the balance of evidence. For example, the discovery that the Irish and American *Sisyrinchium bermudiana* is different from the Greenland representatives of the genus refutes the suggestion that *S. bermudiana* might have been brought to Ireland from America via Greenland by migrating geese.[223]

Two similar taxa occupying separate geographical (or ecological) areas are known as *vicariants* or *vicariads*, and the phenomenon as *vicariance* or *vicariism*. As may be seen above, the difference between vicariance and disjunction is one of degree only. Thus the genus *Cedrus* is a disjunct taxon, occurring in four separate areas: Atlas mountains of Morocco and Algeria; Cyprus; Lebanon, Syria and south-eastern Turkey; and western Himalayas. But at the specific level it provides an example of vicariance, for the four areas are occupied by four separate species: *C. atlantica*, *C. brevifolia*, *C. libani* and *C. deodara* respectively. Like disjunct taxa, vicariants may arise in various ways: by a taxon migrating to a new area and evolving there into a different taxon; by a formerly continuously distributed taxon becoming separated into different areas and there undergoing divergent evolution; by parallel evolution of two taxa from a common ancestor in two separate areas; or by convergent evolution (in response to similar environmental conditions) of two taxa from separate ancestors. The nature of the last category (*false vicariance*) can frequently be uncovered by detailed studies, for the similarity of the apparent vicariants is often only of a superficial nature and the taxa concerned are best not termed vicariants. Cutler[85] showed that there are, contrary to prior belief, no genera of Restionaceae common to South Africa and to Australia, although there are several superficially similar genera.

Geographical patterns exhibited by vicariants are often, like those of disjuncts, recurrent, and indeed often coincide with them. There are a number of taxa found in the mountains of southern Spain and those of the eastern Balkan peninsula, but not between these areas. The montane Spanish *Viola cazorlensis* is represented by a closely similar vicariant, *V. delphinantha*, in Greece and Bulgaria, while *Convolvulus lanuginosus* from southern France and Spain and *C. calvertii* from the Crimea and south-western Asia represent vicariants in lowland habitats. *Convolvulus boissieri* appears indistinguishable in southern Spain and the eastern Balkans and thus represents a disjunction rather than a vicariance, but involving the same areas and with presumably the same underlying causes.

Vicariance biogeography

The phenomena of vicariance and disjunction have in recent years been investigated with renewed enthusiasm, utilizing the techniques and principles of cladistic analysis (see Chapter 2) to interpret distribution patterns.[76,304,476,477] This approach is called *cladistic biogeography* or *vicariance biogeography*, in which the word 'vicariance' is often used to describe the *formation* of vicariant patterns (i.e. physical separation of populations followed by allopatric speciation), rather than to refer to the *state* of vicariism. (This dual use of the word 'vicariance', however, remains widespread.) By this method, of which there are many variants, cladograms of taxa are constructed (Fig. 7.3A) and the names of taxa at the branch-ends are substituted by the areas in which they occur, forming a so-called *area-cladogram* (Fig. 7.3B). If a pattern can be repeatedly reconstructed using different groups of organisms, then there is reasonable evidence to believe that that pattern is a true representation of the relative origins of the floras (and faunas) of the areas concerned. To aid in the visual grasp of the data, the area cladogram can be represented on a map, with the areas (in their correct places) linked as in the cladogram, the lines being (misleadingly) called *tracks* (Fig. 7.3C). If the similar individual tracks

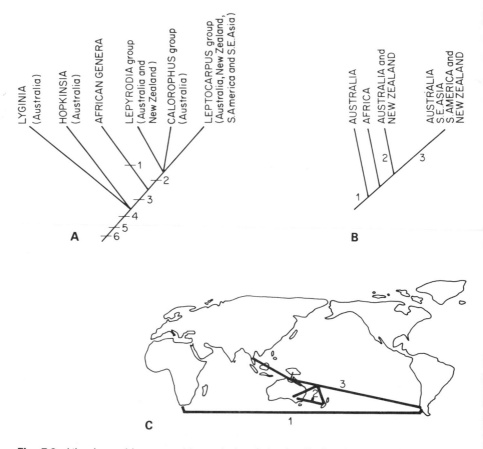

Fig. 7.3 Vicariance biogeographic analysis of the family Restionaceae, redrawn from Linder[261]: **A**, cladogram of the family (numbers represent character-state transformations); **B**, area cladogram derived from the latter; **C**, cartographic representation of the area cladogram, showing tracks.

of a wide range of organisms are superimposed, a ***generalized track*** depicting the summation of the individual tracks can be constructed.

It is the interpretation of these generalized tracks that is the most contentious aspect of vicariance biogeography. It is argued that if widely different groups of organisms, for example animals and plants with different methods and efficiencies of dispersal, conform to the same generalized track, then disjunctions in the areas of distribution are not likely to have arisen from dispersal across barriers, nor to extinctions from intermediate areas. The alternative explanation championed is that of a vicariance event—the disjunctions being erected by geological events (such as the formation of intervening seas or mountain ranges), rather than by migrations or extinctions of the organisms themselves. For certain recurrent patterns of disjunction, particularly those involving South America, Africa and Australasia, the geological evidence for vicariance events is now extremely strong. The idea of ***continental drift*** was first forwarded by the Austrian A. Wegener in 1912, but generally accepted only in the 1960s. It is now known that the earth's crust is made up of a series of plates, on which the land-masses are based, and that these plates have moved and are still moving in relation to one another. In particular the southern hemisphere land-masses mentioned above formed one southern super-continent at the most about 200 million years ago. Their separation and the formation of the southern oceans are still continuing today; for example Africa and South America started to separate about 180 million years ago, and Australia from Antarctica only about 55 million years ago. India also formed part of this southern continent; it broke finally from Madagascar about 100 million years ago and moved northwards to collide with Asia (with an impact causing the formation of the Himalayas) about 55 million years ago. It is not surprising that such catastrophic events took half a century before they were accepted as fact, but the geological evidence is now virtually unequivocal and application of these facts to biogeography offers plausible explanations of many hitherto mysterious puzzles in distribution.

Hence the subject of vicariance biogeography grew up in the 1970s, soon after the idea of continental drift became firmly accepted. It has been and is still studied mainly by workers interested in world-wide distribution patterns ('panbiogeography') rather than by those interested in small-scale problems of plant distribution. Thus there has been an over-emphasis of the significance of continental drift at the expense of other vicariance events, and indeed on the importance of vicariance events at the expense of dispersal events. We know of a great many cases where dispersal must have given rise to the present disjunctions or vicariisms, and there have not been sufficient detailed studies for us to be able to assess the relative importance of dispersal and vicariance events as disjunctive phenomena. To claim that either one is vastly more important than the other, as many do today, is premature and unscientific. Instead, it is clear that both are highly significant processes and that both must be borne in mind as possibilities when investigating a fresh distribution pattern. Nor is it certain that the two processes are always so different. Geographic disjunctions mostly do not arise overnight, but gradually, and there is no reason to believe that plants did not migrate to fresh areas at about the same time that these areas were becoming isolated. The Post-glacial recolonization of Britain from continental Europe might well come in this category. Indeed, Rose[365] believes that some of the rare plants of south-eastern England arrived there from continental Europe by vicariance events and others by dispersal events (i.e. before and after formation of the English Channel respectively).

To summarize, one may conclude that the strengths of vicariance biogeography are the use of cladistic analysis to define monophyletic groups and to demonstrate their past relationships *before* the geographical interpretations are made. It becomes especially powerful when the area cladograms and generalized tracks can be correlated with known and accurately dated geological history. Its weaknesses are the weakness of cladistics (often analysis does *not* succeed in revealing monophyletic groups and their relationships) and the over-inflated claims of some of its disciples.

Endemism

Taxa which occur only in single restricted geographical areas are known as *endemics*. Endemism again is a relative concept but is normally applied only where there is a considerable restriction in the area of distribution. It should be noted that in biogeography the term endemic is used in a different sense from its everyday use (especially in the description of diseases), which is exactly equivalent to the meaning of the biogeographic term *indigenous* (native). There are basically two types of endemism—*neoendemism*, where a taxon is evolutionarily young and has not yet been able to spread to other areas, and *palaeoendemism*, where a taxon is now restricted but once enjoyed a far wider distribution. Examples of the former are *Senecio cambrensis*, a new amphidiploid which arose in N. Wales this century and is still endemic to Britain, and *Crepis fuliginosa*, which arose by chromosomal rearrangements in the Balkan peninsula from its more widespread ancestor *C. neglecta*. Examples of the latter type are *Ginkgo biloba*, now restricted to a small part of China but widespread as a fossil in the North Temperate zone, and *Sequoiadendron giganteum*, formerly similarly widely distributed but now found only in the Californian Sierra Nevada. Richardson[359] referred to endemics intermediate between the above two extremes, i.e. plants which are not of recent origin but have retained a narrow distribution, as *holoendemics*. In addition, local conditions may sometimes induce the 'reactivation' of palaeoendemics, which evolve new endemics after a long period of range contraction. Such endemics have been termed *active epibiotics*, and examples from the Canary Islands are given by Bramwell.[47]

Favarger and Contandriopoulis[132] have made extensive studies of the endemic angiosperms of the mountain ranges of western Europe. Using cytotaxonomic criteria they recognized three different types of neoendemics: *schizoendemics*, derived from or having given rise to a more widespread taxon with the same chromosome number; *patroendemics*, restricted diploids which have given rise to more widespread polyploids; and *apoendemics*, restricted polyploids which have arisen from more widespread diploids.

Since neoendemics greatly outnumber palaeoendemics the percentage of endemic species in a flora is usually proportional to the degree of isolation of the area involved. The degree of isolation may be measured either as the *distance* from other similar areas or the *length of time* that the area has been isolated; often both these factors apply. Figures for endemics given by Favarger and Contandriopoulis for various mountain areas (and referring *only* to exclusively montane species) are: Corsica 38%, Balkans 37%, S.E. Spain 36%, Eastern Alps 18%, Pyrenees 14% and Western Alps 13%. The British Isles are separated from France by approximately 35 km, and have been so for only about 7 500 years. Accordingly, they exhibit about 1% endemism (19 non-agamospermous species),[462] only about a third of that of the Balearics, which have a slightly smaller total flora but a greater degree of and a more ancient isolation. The Canary Islands, with fewer native species still but an even longer history of isolation, possess about 47% endemics.[47] In greater contrast, remote oceanic islands or archipelagos such as Hawaii have a very high degree of endemism (often 80–90%).

Larger areas usually have a greater proportion of endemics than smaller areas, for obvious reasons; for the whole world the figure is 100%! Thus the numbers of species endemic to the Eastern Alps *or* to the Western Alps added together amount to only 78% of the total number of Alpic endemics, the other 22% being species found in both the Eastern and Western Alps. But when one is comparing a small isolated area with a large relatively uniform area the reverse is often true. Whereas the mountains of S.E. Spain exhibit 36% endemism, the whole of Spain possesses approximately 20% endemics[201] and the whole of Europe about 33%.[469] Useful data on endemism in California are given by Stebbins and Major.[423]

Obviously the taxonomic treatment of a flora greatly affects the statistics concerning its endemism. None of the endemic British species listed by Walters[462] is taxonomically remote from a non-endemic species, and an ultra-conservative taxonomist might unite all the endemic species with non-endemic ones, thus reducing the endemism to 0%. At the opposite extreme, Walters included only four agamospecies in his figures. There are in fact several hundred agamospecies which are endemic to the British Isles, including over 200 in the genus *Rubus*.[307]

Centres of diversity

If the distribution of every species within a genus is drawn on a single map, it will usually be found that there are one or more areas with a marked concentration of species. Such an area is termed a *centre of genetic diversity* for that genus. Often there is a single such area for the genus (or other greater or lesser taxon) concerned, and there are progressively fewer species found as the distance from the centre of diversity is increased. This topic was first studied in detail for crop plants, particularly cereals and legumes, by Vavilov in the 1920s and 1930s,[454] and an updated discussion of the topic is provided by Zeven and Zhukovsky.[486] Vavilov and later workers found, moreover, that there is a frequent coincidence in the centres of diversity for many different, unrelated taxa, and that a relatively small number of major centres of diversity can be recognized in the world. Zeven and Zhukovsky defined twelve such centres (Fig. 7.4), but there are good arguments for recognizing fewer and smaller centres. All of them lie in tropical, subtropical or warm-temperate areas, and they have provided virtually all the world's important crop plants. When the centres of genetic diversity for non-crop

Fig. 7.4 Zhukovsky's map of the twleve centres of genetic diversity, adapted from Zeven and Zhukovsky.[486] Superimposed on this map (in black) are the three major centres in which agriculture is believed to have begun (after Harlan[184]).

species are similarly plotted, they are mostly found to coincide with those of the crop species.

Vavilov believed that the centres of diversity were also the centres of origin of the taxa concerned. This may well be true in some cases, especially in the tropical parts of large continental masses which have probably varied in climate rather little for many thousands of years, but in other areas there have been major changes in climate and large-scale movements of land-masses since the time of origin of modern plant taxa, and it is unlikely that there is in general a close relationship between the centres of diversity and origin. Nevertheless, the centres of diversity today probably hold the main remnants of the ancestral genetic material of the taxa concerned. The older the taxon, the less likely the coincidence of the centres of origin and diversity. Stebbins,[421] however, considered that the processes of evolution have operated in the past in a similar manner to now ('principle of genetic uniformitarianism'). Using this principle he deduced that angiosperms originally diversified in semi-arid zones, since those habitats now support the greatest number of species per genus.

Each monophyletic taxon has, by definition, a single centre of origin. By study of the characteristics of the taxon in its centre of diversity in relation to its characteristics progressively further from the centre, the major patterns of evolution and pathways of migration can very often be deduced. The ways in which the taxon has adapted to the different environments which it has encountered during its migrations are thus frequently uncovered. Such an 'outward' pattern of evolutionary migration is known as *adaptive radiation*. Its demonstration may be of great assistance in deciding upon the taxonomic relationships and delimitations of the various biotypes encountered within a taxon. As precise as possible a characterization of the biotypes is important, since anything less might well fail to differentiate between separate biotypes which are superficially similar. Frequently cytological study is important. It is commonly found, especially in temperate taxa, that plants near the centre of diversity are diploid, whereas high proportions are polyploid towards the edge of the range, i.e. polyploidy appears to have been an important aid to adaptive radiation in many instances.

It should be emphasized that the methods described above are just as often valuable in studying the migration of the variants of a single species as in studying the species of a single genus. Favarger[131] has found that the cytology of a great many alpine endemics throws light on their evolutionary pathways. For example, the fact that *Paradisea liliastrum* is diploid in the Pyrenees but polyploid in the Alps suggests that it migrated from the Pyrenees to the Alps, where it became polyploid.

Sometimes there appears to be more than one centre of genetic diversity for a taxon. This might indicate that the taxon is polyphyletic, having arisen in more than one area. However, often all but one (or even all) of the centres are **secondary centres of genetic diversity**—areas where the taxon was able to diversify (perhaps due to the presence of a large number of habitats) to a greater extent than in other areas a similar distance from the primary centre. For example, a primarily South African taxon might exhibit a secondary centre of diversity in the Mediterranean basin, or a primarily Balkan taxon might have a secondary centre in Spain.

Usually it is possible to differentiate between primary and secondary centres of diversity if thorough biosystematic investigations have been carried out. Thus the grass genus *Vulpia* has its primary centre of diversity in the western Mediterranean, with secondary centres in temperate North and South America and in south-western Asia. The diversification in the latter areas is based on only one of the five sections of the genus found in the western Mediterranean and, moreover, there are higher proportions of polyploid taxa in the secondary areas than in the primary one.

This decrease in variability, i.e. in the number of biotypes, away from the centre of diversity is known as **biotype depletion**, which, as intimated above, can be used to investigate pathways of migration. Böcher[41] found that *Ranunculus glacialis*, an arctic-alpine species, is far richer in biotypes in the Alps than in the arctic. This diminution might have originated either during a northwards migration after glaciation or from a greater loss of biotypes in the north during perglacial survival. Biotype depletion is also one possible explanation for the observation that species frequently appear to be far more demanding ecologically near the edge of their geographical range than in the centre of it. In Europe this is very often manifested by the northern botanist finding that species which are confined to particular habitats in his own country are far less so restricted in southern Europe.

The concept of centres of genetic diversity (and, even more, centres of origin) has been strongly criticized in recent years by vicariance biogeographers,[76] who prefer to think in terms of the 'generalized track' of a taxon differentiating allopatrically in different regions. Almost certainly, this is a better model than a centre of origin in some instances, a less good one in others, and in yet further cases something between the two is most accurate. As intimated earlier, a centre of origin is a more valid concept in the case of younger taxa (and hence usually at the lower levels of the hierarchy), particularly where differentiation has occurred more recently than any known vicariance events. The evolution of modern cereals is a good example.

Ecological differentiation

The ecotype is defined in Chapter 1. For present purposes it can be considered as a biotype (or group of biotypes) which is adapted to a particular ecological niche, but which is fully interfertile with other ecotypes in the same ecospecies.

Ecotypification, the formation of ecotypes in different habitats, has been the subject of many detailed classical studies. Best known are the works of Turesson,[440] who investigated the 'genotypical responses' of a number of common species from different habitats, such as woodland, open fields and sand-dunes, in south-western Sweden, and of Clausen, Keck and Hiesey,[65,66,67] who studied the extent to which a range of plants are structurally and physiologically adapted to different altitudinal zones from the Pacific coast to high altitudes in the Sierra Nevada of California. These workers have found that very many plant species have adapted to the different conditions which they encounter over their total geographical range. Ecotypification does not imply merely an acclimatization, akin to 'hardening-off' plants taken from the glasshouse to be placed in the cooler conditions outside, but involves the evolution of distinct genetic races. Thus, for many purposes, the specific name of the plant is not sufficient to convey precisely the nature of the plant in question. Whether ecotypes (and other infraspecific variants) should receive scientific names is a matter for debate, but several workers have argued that they should.[389,406,448]

Some of the most revealing studies on ecotypification have been carried out on the differentiation of populations on different islands,[155,427] for islands provide a ready-made experiment in which the discreteness of the natural populations is assured (Fig. 7.5).

The extent to which ecotypes have become morphologically differentiated varies enormously. Sometimes morphological adaptation is the most crucial factor, for example in dwarfer alpine ecotypes or larger-leaved, less hairy shade ecotypes. In other ecotypes no exomorphic changes are visible. Examples of chemical ecotypes are given in Chapter 4. Stace[406] listed cases of physiological ecotypes involving adaptations to differing soil conditions (salinity, pH, heavy metals, nutrients, moisture), climate (day-length, temperature, length of seasons, shade) and biotic influences (grazing, parasitism, cultivation).

As mentioned previously, there is a danger that variants with a similar appearance might be automatically referred to the same biotype. This is particularly a problem with classical taxonomic work, since one might be relying on a written description or at best a dead specimen. There must be many cases like this where, say, a dwarf variant is automatically referred to 'var. *nana*', although the dwarfness is only one manifestation of the biotype. The subjective selection of this single character results in an artificial classification at the infraspecific level. Convergence in major exomorphic features between only distantly related plants is commonplace, for example the cactus-habit in Madagascan *Euphorbia* species and the *Cuscuta*-like habit in the Lauraceous *Cassytha*. It would be most surprising if such convergence were not equally prevalent between infraspecific variants, and indeed it has frequently been shown to exist, for example in biotypes of different species adapted to growth on serpentine rocks.[253]

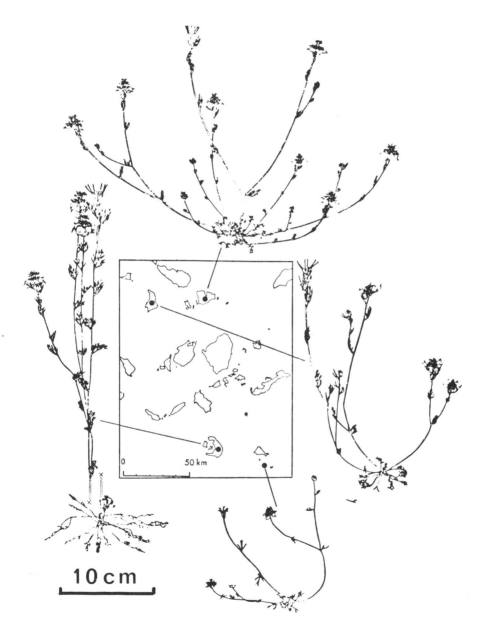

Fig. 7.5 Genetic variation in growth-habit of *Nigella degenii* from different Aegean islands, taken from Strid.[427]

Ecotypes are discrete biotypes characteristic of distinct habitats. Where the habitats are not distinct, but form a continuous range, the ecologically adapted species also exhibit a continuous range of variation (Fig. 7.6). Such a range of variation is termed a ***cline***, of which two main types can be distinguished: ***ecoclines*** (clines along an ecological gradient) and ***topoclines*** (clines along a geographical gradient). Clines can exist over very short or very great distances. Gregor[166] found that coastal populations of *Plantago maritima* not only exhibit an ecocline in growth-habit from low-lying waterlogged salt-marshes to the higher, well-drained areas landwards, but also a much broader topocline in scape length from western North America eastwards to Central Europe (Fig. 7.7). In tropical climates clines are common from equatorial regions northwards or southwards to warm temperate regions, or even in both directions. In *Combretum molle*, for example, plants from equatorial regions have large, thin, sparsely hairy leaves and grade northwards to Arabia and southwards to South Africa into plants with small, thick, densely hairy leaves (among other characters).

Although earlier workers tended to stress the importance of either ecotypic or ecoclinal variation, according to the results of their own researches, there is every sort of situation in nature from one extreme to the other.[46] Where clines are quite continuous the taxonomic problem posed is insoluble, but

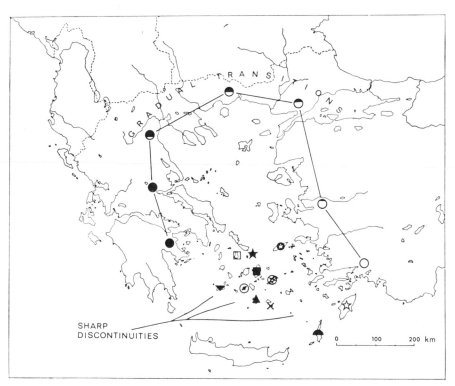

Fig. 7.6 Diagrammatic representation of variation patterns in the *Nigella arvensis* complex in the Aegean, taken from Strid,[427] illustrating the sharp discontinuities between island populations in contrast to the clinal variation in mainland areas.

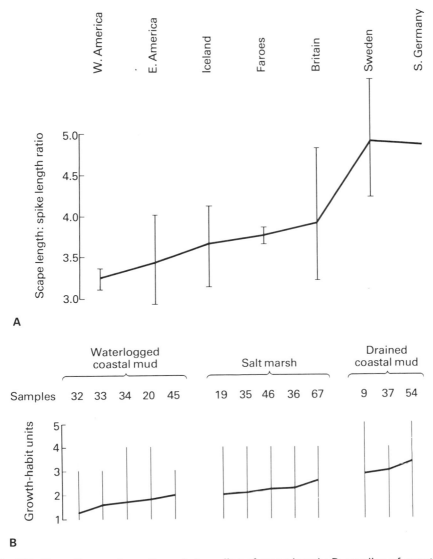

Fig. 7.7 Topoclines and ecoclines: **A**, topocline of scape length; **B**, ecocline of growth-habit, both in *Plantago maritima*, adapted from Gregor.[166]

fortunately there are usually gaps in the cline, caused by jumps in the environmental gradient, which enable taxa (often subspecies) to be delimited.[112]

Clines may exist not only in the expression of the characters of individual plants along a gradient, but in the proportion of plants bearing different characters, i.e. in polymorphisms. A *polymorphism* is the presence in a population of different attributes of a character such that the proportion of plants with the rarer attribute is greater than would be expected from random mutation.[220] Two well-known cases also provide examples of clines. In

Spergula arvensis there are two main variants, var. *arvensis* with papillate seeds and var. *sativa* with broadly winged seeds. The proportion of these two in populations varies in Great Britain in a topocline running from southern England (with up to 90% var. *arvensis*) to northern Scotland (with 0% var. *arvensis*).[305] In *Trifolium repens*, Daday[88] found that there is a cline in the proportion of cyanogenic (prussic acid releasing) plants across Europe, with most in the south-west and least in the north-east.

Polymorphisms by no means always form clines; the proportion of each polymorphic form (morph) may remain fairly constant or vary in a regular or irregular pattern from population to population. Much research has been carried out on the adaptive significance of polymorphic variation, but taxonomists have paid little attention to the subject.[447]

All the examples given above have been taken from vascular plants, for the study of ecological differentiation in lower plants has only just begun in earnest. The few results so far obtained indicate that bryophytes and algae will provide plenty of examples of all the above phenomena.

Presentation of ecological data

Ecotypic differentiation occurs at every taxonomic level from the species downwards, the level being decided by the other characters with which it is correlated. Ecological characters in general may serve to reinforce taxonomic separation at far higher levels, for example fresh-water and salt-water algae, tropical and temperate families of vascular plants, and epiphytic and terrestrial genera of orchids.

Floras generally include a statement on the ecological preferences of the taxa described. Some recent British local Floras, for example that of Cadbury, Hawkes and Readett,[57] whose data have been gathered systematically (often on specially designed record-cards) over a fixed period of time, have analysed the ecological information obtained and presented it, sometimes in a codified form, consistently for each taxon. Thus precise statements on the proportion of occasions in which the taxon was recorded in each habitat-type have replaced subjective observations on habitat preferences.

Such analysis is one aspect of the subject of **phytosociology**, the description and classification of vegetation-types according to their constituent species. Most phytosociological work has been carried out in continental Europe, based on the Zurich–Montpellier School of research which arose in the last quarter of the nineteenth century and later became popularized mainly by the work of J. Braun-Blanquet and R. Tüxen. This work expresses the data obtained in a hierarchical classification system akin to the taxonomic hierarchy (Table 7.1). To some taxonomists this seems a cumbersome and inflexible method of expressing ecological data, but if the units named are indeed entities consistently recognizable in nature, then they do form a realistic method of summarizing the ecological characteristics of taxa. Recently they have been utilized to indicate ecological preferences in the new *Flore de France*.[170] If this method proves useful their appearance in future Floras and monographs may be expected.

Snaydon[388] has argued that greater use should be made of ecological characters in the *construction* of classifications, rather than in presenting

Table 7.1   The accepted phytosociological ranks, taken from Shimwell.[377]

Rank	Ending	Example
Class	-etea	Festuco-Brometea
Order	-etalia	Brometalia erecti
Alliance	-ion	Mesobromion
Association	-etum	Cirsio-Brometum
Sub-association	-etosum	Brometum caricetosum

ecological data simply as supplementary information, and that continuous variation should be taken into account along with discontinuous variation. It is probably true that taxonomists have fought shy of ecological data largely because they are so often of a continuous nature. Guinochet[169] considered that phytosociological data can indeed be used in the process of classification, since he believes that species which are constituents of several different plant associations are represented in each of the latter by different genotypes. But, as Heywood[203] showed, new approaches to the presentation of ecological information and a reappraisal of infraspecific categories need to be made before ecology can become a major direct tool in plant taxonomy.

Alien plants

A proportion of the flora of every region consists not of *native* plants, but of *alien* species from other areas introduced deliberately or accidentally by man. The terminology which has been suggested for such **anthropochores** (man-distributed plants) is complex and varied, being based upon their grades of establishment and their methods of arrival, and need not be gone into here.[262] For practical purposes alien species may be divided into **naturalized aliens** and *casuals*—the former maintaining themselves once they are introduced but the latter dying out quickly due to adverse conditions. There is, of course, every grade of intermediate situation.

The proportion of alien species in different floras varies a great deal. In undisturbed regions they are virtually absent, but in areas with poor native floras and extensive human settlement, for example much of north-western Europe, they form the bulk of the flora in terms of numbers of species. In the British Isles there are about 1 550 native species of vascular plants (excluding agamospecies), but at least 5 000 alien species have been recorded (E. J. Clement, pers. comm.). Most of these are casuals, but there are at least 800 naturalized aliens. In the whole of Europe, Webb[469] has calculated that there are about 11 000 native species and about 900 naturalized aliens, but his criteria for the acceptance of naturalized species are very strict. Alien species of algae and bryophytes appear to be much less prevalent in all floras, although there are well-known examples, e.g. at least 13 species of bryophyte in the British Isles.[79]

Even in relatively unexploited tropical regions aliens may be abundant. In Tropical Africa most of the crop plants and urban ornamental trees are aliens. In tropical areas in general there are a great many very widely distributed weeds whose area of origin is unknown or uncertain. The amphidiploid *Poa annua* is an almost cosmopolitan weed, occurring from the arctic to the antarctic, but judging from the distribution of its putative parental ancestors it arose in the western Mediterranean region.

Many aliens have become introduced by highly circuitous routes. *Trifolium subterraneum* was introduced from Europe to Australia by the early settlers, probably in sheep's wool, but in recent years it has been re-introduced into Britain in untreated fleeces. The reintroduced plant differs from the native British stock, being much more robust and frost-sensitive, and in some areas of Britain it has become the more common variant.

Alien species are an *important* part of a flora for four main reasons. Firstly, if their origin remains undiscovered, they will lead to false results in floristic analyses (see under Disjunction, p. 160). Secondly, if their presence is undetected, they will be mis-identified and lead to wrongly documented experimental and observational results. *Epilobium ciliatum*, an alien from North America, had occurred in Britain for about forty years before its discovery in 1932, during which time it was 'identified' as various different native taxa. Thirdly, alien species often hybridize with native ones, and in some cases new successful amphidiploids may result. *Senecio squalidus* and *Spartina alterniflora* are two such aliens in Britain.[405] Fourthly, alien species often have serious (not always harmful) ecological consequences, both in natural vegetation and in cultivated areas, for example *Opuntia* in Australia, *Acer pseudoplatanus* in England and *Eichhornia* in Africa.

Phenotypic plasticity

The familiar equation *Genotype + Environment → Phenotype* implies that the phenotype is not merely a manifestation of the genotype, but that environmental factors play a part in modifying the latter to produce the former. As pointed out in Chapter 1, Turesson referred to different phenotypes which were merely the product of differing environments as ecophenes, and Clements called them ecads. The ability to express a genotype as different phenotypes according to external conditions is referred to as **phenotypic plasticity**, or one may refer to **plastic responses**. It is important for taxonomists to recognize plasticity and to decide whether or not the different ecophenes should be taxonomically named.

Work on plasticity formed some of the earliest biosystematic experiments. In 1902 J. Massart showed that the different growth-forms of *Polygonum amphibium* were adaptations depending upon whether the plant was growing in water or on dry land (or in intermediate conditions), as they could be readily interconverted by transplanting. G. Bonnier carried out a wide range of cultivations, starting in 1884, in Paris, the Alps and the Pyrenees, and claimed to demonstrate that a number of species could be interconverted according to the altitude at which they were grown, for example *Lotus alpinus* and *L. corniculatus*. However, some of Bonnier's conclusions are now known to be ill-based, the apparent convergence in characters of separate species often involving superficial features only, or being based on faulty experimental technique. Nevertheless, more sophisticated experiments along similar lines were later carried out by A. Kerner and by F. E. Clements, the latter leading to the long-term studies of Clausen, Keck and Hiesey.[65,66,67]

Certain environments are well known to give rise to extreme ecophenes in a wide range of species, for example adaptations to shady/sunny, alpine/ lowland and wet/dry conditions. Certain genera or species are also notorious for producing a wide range of ecophenes, and the taxonomist has to be on the look-out for them. In *Epilobium* sun-plants have small, thick leaves, much anthocyanin, many hairs, and a short stature, whereas shade-plants have the opposite characters. Since one of the most important diagnostic characters in *Epilobium* is the type of indumentum, it is vital to examine the *quality* of the hairs, not their quantity, when making determinations. In many annual grasses and other plants dry conditions promote a dwarf habit, a normally tall plant with a well branched inflorescence often appearing as a midget a few centimetres high with a single spikelet or flower (Fig. 7.8). Fortunately the measurements of the individual parts of the latter usually are not or are much less modified.

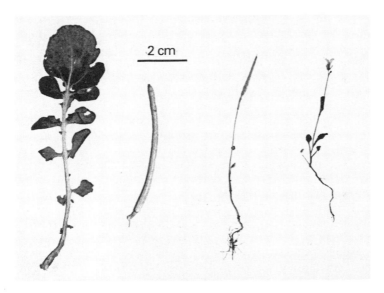

Fig. 7.8 Photographs of specimens of *Barbarea verna* collected from the same locality. One leaf and one fruit from a well-grown specimen 90 cm high and bearing about 100 flowers are shown together with two dwarf specimens each bearing only one flower or fruit. The dwarfing is due to poor, dry soil conditions.

Plasticity is extremely common in bryophytes, for example the genera *Hypnum*, *Sphagnum* and *Scapania*, where a great number of taxa have been described based on them. In the 1920s H. Buch pioneered the study of plasticity in bryophytes, conducting cultivation experiments on a range of liverwort genera. Many examples are known also in marine, fresh-water and terrestrial algae and in lichens; in the latter case the nature of the substrate (e.g. rock or tree-trunk) can be the important factor.

Plastic responses are by no means confined to exomorphology, for chemical and anatomical characters are frequently affected. Several examples of the former are given in Chapter 4, and numerous cases of plasticity in leaf epidermal characters have been documented by Stace.[402] Cook[70] has shown that the ecophenetic response is complex and varied, and that similar responses may have quite different causes and mechanisms.

Certain genera of flowering plants adopt different phenotypes according to the time of year at which they germinate and flower—the so-called *seasonal variants*. This phenomenon, often referred to as *pseudoseasonal polymorphism*, is particulary notable in montane taxa, which may be subjected to different lengths of season and temperature regimes according to the altitude at which they grow and the agricultural practices to which they are subjected; *Gentianella*, *Melampyrum* and *Rhinanthus* are good examples. Since these variants are often separated spatially they have mostly been given different names, frequently at the species level. In modern Floras they are often treated as varieties or subspecies, but in most cases they have not been investigated adequately (the three above are difficult to cultivate) and are often probably just ecophenes.

It has been pointed out by several workers that plasticity and ecotypification are alternative adaptive strategies which are both important in evolution. Even where variation is found to be entirely ecophenetic, it has sometimes been demonstrated that the *extent* of plasticity is genetically controlled. Schlichting and Levin[373] found that in several groups of species the degree of ecotypic variation and the extent of plasticity possible are not or are only very loosely correlated.

Particularly perplexing are cases where certain ecophenes can mimic genuine ecotypes, for example dwarf variants of *Prunella vulgaris* adapted for growth in close-cut lawns,[302] or of *Cytisus scoparius* adapted for growth on exposed maritime shingle.[154] In such examples the variant might be a genetically determined dwarf or a dwarf ecophene, and only experimental cultivation will distinguish the two.

The majority of taxonomists are of the opinion that ecophenes should not be given taxonomic status. There are a great many taxa (mostly infraspecific) based upon ecophenes, and when a taxonomist discovers their background they are usually no longer recognized as distinct, but are relegated as mere synonyms. Inevitably this process of discovery is slow, as most species have not yet been systematically cultivated. Probably the rate of description of new taxa which are in fact ecophenes still exceeds the rate of their relegation to synonymy. This is certainly so where little biosystematic research has been undertaken, particularly in the tropics and in lower plants.

Although in vascular plants there is almost universal acceptance that ecophenes should not be taxonomically named, there are some exceptions; Jones and Newton[236] treated *Puccinellia pseudodistans* as *P. fasciculata* forma *pseudodistans* because their experiments indicated that it is a growth form of *P. fasciculata* induced by waterlogged soils (although others dispute their results). Bryologists realize that the proportion of infraspecific taxa (and even species) which are ecophenes is far higher than in vascular plants, but nowadays there is some agreement that, where detected, they should lose their status.[381] In the algae the situation is more difficult still, because of the lack of experimental evidence. The lichens are of interest since lichenologists have for many years *deliberately* given names to ecophenes, usually at the level of the forma, although even in this group such a practice is being abandoned.[327]

Table 7.2 The categories proposed by Buch[55] for the recognition of ecophenetic variants of liverworts.

	Modification	Description and cause
1.	*parvifolia*	Small leaves; in faint light
2a.	*laxifolia*	Distant leaves; in faint light and moist air
2b.	*densifolia*	Dense leaves; in dry air
3a.	*leptoderma*	Thin cell-walls; in very moist air or water
3b.	*pachyderma*	Thick cell-walls; in dry air
4a.	*viridis*	Green, cell-walls colourless; in diffuse light
4b.	*colorata*	Red, brown or purple cell-walls, obscuring the chlorophyll; in direct sun-light

Combinations, e.g. mod. *pachyderma-viridis*, may also be used.

Ecophenes are entities of a quite different nature from genetically determined variants, but they can be conspicuous in nature and they are of ecological importance. On many occasions it can be advantageous to refer to them concisely, i.e. by means of a name, and the system developed by Buch[55] appears to be very suitable for this purpose. In this system (Table 7.2) each ecophene is referred to as a ***modification*** (abbreviation ***mod.***) and those characteristic of each habitat receive the same name. There is, therefore, no question of any recourse to a code of nomenclature, or to adopting the earliest available name, as the selection of names is automatic. Buch's system was developed for liverworts; for other groups it would need extension, or perhaps replacement by a more appropriate series of names, and the ambiguous term 'modification' would also be better replaced by a more precise one.

Section 3
Taxonomy in Practice

Introduction

Once a thorough survey of the patterns of variation exhibited by the taxonomically significant characters of a group of taxa has been made, it should be possible to construct a predictive (natural) classification of that group. This involves arranging the taxa, according to the principles inherent in whatever methodology is adopted, in the hierarchy of taxonomic ranks listed in Chapter 1. In theory this may seem straightforward, but in practice difficulties are often encountered. Frequently these involve such considerations as the nature of the taxonomic characters that should be emphasized (especially when there are apparently conflicting data), the ranks that should be used to express the different levels of relationship (i.e. the absolute as opposed to relative degrees of similarity), and the ways in which taxa showing varying degrees of distinctness (often owing to isolation by different factors for differing periods of time) should be treated. This is the subject matter of Chapter 8.

Chapter 9 considers the physical resources which taxonomists utilize in making classifications. Finally, Chapter 10 discusses the priorities in plant taxonomy today, and attempts to show the vital role which it must play in improving the welfare of mankind.

8
The process of classification

According to Adansonian principles all data should be used in constructing a classification without preference for one sort over another, i.e. without weighting (the *empirical* approach). In practice the great majority of taxonomists are forced, by strictures of time and manpower, to be selective in their use of characters, and much discussion on the most valuable sorts of characters still takes place.

The early taxonomists believed that some characters, the so-called *essential characters*, which defined the *essence* or essential nature of the organism, were taxonomically the most important. These were detected by *deductive* (*a priori*) reasoning, often by reference to important or essential functions which the organs or characters performed. The flowers of angiosperms, for example, were considered essential structures and they provided a high proportion of the characters used in classifying these plants. This sort of taxonomic approach was particularly clearly propounded by botanists such as Caesalpino and de Candolle, but dates back to the days of Theophrastus. Cain[58] described de Candolle as a 'brilliant late representative' of the deductive school, but he also showed that de Candolle realized the limitations of the approach and Cain concluded that de Candolle's system was in fact transitional from entirely deductive systems to a purely empirical one.

Proponents of the cladistic approach (see Chapter 2) use a different method to select the most important characters, seeking only advanced (apomorphous) character-states that are found to be correlated with one another and that are not the result of convergence, parallelism or reversal.

Functional characters

Since the abandonment of the search for essential characters there has been much argument as to whether obviously *functional characters*, which therefore have been and are exposed to strong selection during evolution, are more or less important taxonomically than non-functional (or apparently non-functional) ones. The subject has been discussed by zoologists more than by botanists. Thus, of ten papers contributed to a symposium entitled 'Function and Taxonomic Importance', organized by the Systematics Association in 1957, only one was by a botanist.[199]

In the twentieth century most plant taxonomists have avoided the use of characters which have obvious and well-defined functions, on the grounds that they have become channelled by evolutionary processes. This channelling must have taken the form of convergence, parallelism and extremely rapid divergence in many cases, so that the possession of such characters in common (or the reverse) is not a very good measure of overall similarity. Succulence, the presence of vessels in pteridophytes and gymnosperms, lack or fusion of petals, and explosive fruits are obvious examples. The case for considering this sort of feature to have little taxonomic value has been put very strongly by Wilmott,[481] among others. It is, of course, a view which is particularly relevant to taxonomists attempting to construct phylogenetic classifications, and less so to ones aiming at a phenetic system. Nevertheless, it is still a view upheld by a great many plant taxonomists, despite attempts by several recent workers to show that there is no *a priori* basis for ascribing a greater or lesser value to functional characters. Thus, there is little evidence for the view stated by Smith[386] that characters *without* obvious function 'are regarded suspiciously by taxonomists'; rather, the reverse is true.

It is dangerous to pay too much attention to the apparent functional significance of taxonomic characters, especially as often we do not know whether or not the features in question are adaptive. But, equally, a character which has clearly evolved more than once in response to a particular evolutionary pressure must be accepted as of limited value in indicating either phenetic or phyletic relationships. The elimination of parallelisms and hence of at least some highly functional characters is a feature of compatibility analysis (p. 59).

Conservative characters

Although flowers are no longer regarded as 'essential' and therefore taxonomically particularly important, they still provide the bulk of information contained in the diagnoses of angiosperm taxa. This is because in general the flowers appear to be more conservative than do most other organs. It is also often stated, with some justification, that endomorphic vegetative characters are more conservative than exomorphic ones, and the same is frequently stated for many chemical characters. *Conservative organs* or *characters* are those which tend to remain relatively unchanged over a long period of evolutionary development, and hence tend to vary little between closely related taxa. Conservative characters are therefore most useful in delimiting the higher taxa, where the emphasis is on the recognition of similarity between the members of a taxon. This is true whether one is consciously constructing a phyletic *or* a phenetic system.

At the lowest levels of the hierarchy (species and below) the emphasis is more often on the distinction between taxa rather than the clustering of taxa into larger ones. Here, therefore, non-conservative characters, i.e. those which show the greatest diversification, are often more useful. There are a great many exceptions to these generalizations, but they are useful 'rules of thumb'.

It must be emphasized that in most cases it is not possible to conclude that a particular character is always more or less conservative than another—in one group of plants it might be more so, in another less so. Also, a character or organ in which conservatism has been demonstrated in one taxon might be lacking altogether in another. In such instances conservative characters have to be sought anew. In lower plants, for example, in which flowers are lacking and the reproductive organs are often short-lived or even rare, vegetative characters have assumed a far greater prominence in classification than in angiosperms.

Discussion of which characters are most useful at different levels of the hierarchy might be considered to run contrary to the opinion, adhered to in this book, that *all* characters should be examined and utilized in classifying plants. There is, in fact, no contradiction, so long as conclusions on the relative values of different organs or characters are made *a posteriori*, from experience gained after analysis of a very wide range of possibilities. All taxonomists, with hindsight, can point out the most useful characters in a taxon which they have classified. The answer to the frequently posed question 'why are the floral characters those most used in classifying the angiosperms?' is therefore simply that we have found by experience that, since they are relatively conservative, they are *usually* the characters which are the most predictive, i.e. taxa delimited by floral characters are *in general* more natural than those delimited by (say) vegetative or fruit characters. Knowledge of this is useful in classifying organisms where there is insufficient time, material or facilities to cover a very wide range of data.

Good and bad characters

The terms **good** and **bad** are often applied to characters of greater or lesser taxonomic value, but it is clear that there are many criteria which can be relevant in such a discrimination. Thus these two terms are strictly relative in application, and what may be a bad character in one taxon (for example, because of its extreme plasticity) might be a good one in another. Ability to pinpoint these is part of the intuitiveness possessed by an able taxonomist, and comes partly from experience. Taxonomists have to be on their guard to avoid overemphasizing the importance of data which come most forcibly to their notice for quite spurious reasons, for example data in whose investigation they specialize, or data gained by new techiques. The delimitation of the Procaryota and of the betacyanin-containing families are discussed elsewhere in this book; in both cases there is evidence to suggest that the radical reclassifications which were carried out need to be compromised upon in the light of later evidence.

The discovery that lysine synthesis proceeds by quite different pathways in plants (via diamino-pimelic acid) and fungi (via α-amino-adipic acid) led to the proposal by Whitehouse[474] that these two groups of organisms are fundamentally different and should be placed in separate subkingdoms. The slime-moulds (Myxomycetes) cannot synthesize lysine at all and in this respect resemble animals, with which they are placed in Whitehouse's classification. Bacteria (and blue-green algae) possess the same pathway as plants. In fact some traditional fungi (Oomycetes) and some traditional plants (Euglenophyta) possess the lysine pathway characteristic of the opposite group, and a reappraisal of the classification of these lower plants and fungi, which is certainly necessary, clearly requires consideration of a much wider range of characters, for example cell-wall constituents, flagellar structure, carbohydrate reserves, cytoplasmic (particularly golgi-body) ultrastructure and nucleic acids.

Despite complications such as those described above, some characters are consistently more conservative than others and, as already mentioned, are therefore usually of greater value in the higher levels of the hierarchy. It is thus possible to generalize to some degree by considering certain characters to be more likely to vary at, say, the family level than at the species level. For example, the character *ovary position* (superior, inferior, etc.) is frequently a familial characteristic, less often a tribal or generic characteristic, and rarely varies below the genus level. In *Saxifraga* this character varies but is constant within each group of related species (often recognized as sections), and it is possibly true that in the whole of the angiosperms it never varies within a single species (except for monstrosities). Other characters, such as petal size and leaf hairiness, are mostly variable at or below the species level.

Evolutionary patterns

Because of the different selective forces which operate both during the course of evolution and in varied habitats, and the different genetic pools upon which these forces act, there is enormous variation in the patterns of evolution of species (and other taxa). A range of taxa will demonstrate every grade of evolutionary 'maturity', so that examination of a wide spectrum of modern taxa illustrates every stage and mode of evolution. It is, therefore, hardly surprising that the recognition of species, genera, families, etc., is more difficult in some groups than in others, and that different criteria have to be applied according to circumstances.

Two extreme modes of *speciation* (the formation of new species) are illustrated by the general processes which were termed *abrupt* and *gradual speciation* by Valentine.[446] A prerequisite for speciation is the effective isolation of two (or more) elements of a species, each of which becomes a separate species. In gradual speciation this isolation is by external isolating mechanisms, usually spatial ones, and the two isolated fractions become distinct entities by the gradual accumulation of genetic differences due to mutation, recombination and selection. It can be viewed as the adaptation of the fragments of an erstwhile single species to the conditions encountered in different parts of its range. The processes of ecotypification and the evolution of geographical races are rightly considered early stages in gradual speciation. At some time during the process of speciation the subgroups usually evolve a genetic incompatibility, so that they are no longer able to interbreed even if they were to come in contact; at that stage they have become genetically, as well as spatially, isolated.

In abrupt speciation the initial isolation is genetical, often by the evolution of chromosomal differences, particularly chromosome number. The best-known of these processes is allopolyploidy, which involves the simultaneous genetic isolation and phenetic differentiation of a new species, and thus represents an extreme, though very widespread, form of abrupt speciation. Since most well-understood methods of abrupt speciation involve hybridization, abrupt and gradual speciation are indeed opposite extremes, the former usually a sympatric and the latter usually an allopatric process. They are, however, extremes in a spectrum of continuous variation rather than utterly different processes.

These diverse patterns of evolution are manifested in the different relative degrees to which various organs have become differentiated in related taxa. This is simply a restatement of the recurrent theme of Section 2, whereby there is clearly no possibility of predicting the most useful taxonomic characters in a taxon in which they have not been investigated. Davis and Heywood[98] claimed that angiosperm organs tend to evolve in four more or less independent groups—underground parts, aerial shoots, flowers, and fruits. Many examples illustrating this generalization are cited in Section 2, where is it also demonstrated that similar statements could be made for various groups of lower plants as well, but it is dangerous to place too much emphasis on such a precise compartmentalization of organs. Nevertheless, this general concept should be extended equally to non-morphological characters, such as the chromosomes and secondary metabolites. It is obvious from the above summary of gradual and abrupt speciation, that the level (and hence taxonomic value) of the breeding criterion is equally variable. In some groups (especially those in which abrupt speciation has occurred widely) genetic breeding barriers are among the first stages of speciation; in others (especially those in which geographical isolation followed by strong selection pressures on floral organs has taken place) they are among the last. The futility of attempting to redefine plant species (or genera) on their ability to interbreed is thus clear.

Four distinct methods used to deduce phylogenetic pathways have been mentioned in this book: fossil evidence (the only direct method, but rarely available); phenetic data (e.g. cytochrome c sequences, morphological trends as outlined in Chapter 2, chemical and ultrastructural data as discussed below in relation to the Chlorophyta); cladistic analysis (as described in Chapter 2); and biosystematic results (e.g. chromosome rearrangements, interfertility). Except at the lowest levels of the hierarchy (below the species) and in a small number of particularly well studied groups, the available evidence falls well short of that sufficient to construct accurate phylogenetic classifications. Nevertheless, when unequivocal phylogenetic evidence has been obtained its incorporation is strongly advocated providing it produces a more predictive classification. A good example comes from the broad classification of green plants, with particular reference to the green algae (Chlorophyta). Traditionally, algae are separated from higher plants (bryophytes, pteridophytes, seed-plants) at a high level in the hierarchy, for example as the subkingdoms Thallobionta and Embryobionta (see Appendix). It has been known for nearly half a century that the various divisions of algae differ fundamentally in diverse ways, notably in their ultrastructure (especially plastids, cell-walls, and flagella) and chemistry (especially pigments, cell-wall polysaccharides and polyphenolics, and food-storage compounds). In most of these characteristics the Chlorophyta (and Charophyta, if these are separated) much more closely resemble Embryobionta than any other algae, and it is presumed that the higher plants arose from the algoid ancestral stock that also gave rise to the modern algal Chlorophyta. These discoveries have led to the suggestion that the division Chlorophyta should be enlarged to include not only the green algae but also all Embryobionta, so that the accepted classification of the algae into about eleven divisions becomes a classification of all chlorophyllose plants. Whether or not this becomes more widely accepted than hitherto, it seems fairly certain that the Chlorophyta *sensu stricto* (i.e. algae only) is a *paraphyletic* group.

A similar case for the paraphyletic nature of the Procaryota can also be made. According to the endosymbiotic hypothesis,[270] now widely accepted, Eucaryota (animals and plants) ancestrally represent associations (symbioses) between plastid-less, mitochondrion-less organisms and Procaryota (bacteria and/or blue-green algae), just as lichens are fungi with an algal associate. Similarly, the rather anomalous Euglenophyta, which lack a cell-wall yet possess chloroplasts very similar to those of Chlorophyta, are thought to represent the descendants of fungus-Chlorophyta symbioses. The symbiotic nature of Eucaryota, and the probably various origins of their plastids and mitochondria, need to be borne in mind when discussing problems of monophyly at the higher levels of the taxonomic hierarchy.[229]

Whether it is considered desirable or not, classifications of most groups of plants are still patently phenetic. Moreover, due partly to the lack of other evidence, most classifications are based largely or solely on morphological characters. In line with the theoretical stance taken in this book, this over-reliance on exomorphic features constitutes a distortion of the evidence, but such characters have the practical advantage that they are relatively easy to observe and to record. Usually exomorphic characters, if they are broadly based, represent a reasonable sample of the patterns of variation, which can be used to construct a general purpose classification, but the incorporation of evidence from other sources wherever possible is highly desirable. Davis and Heywood[98] have argued strongly for the recognition of species on a primarily morphological–geographical basis. It is, as they put it, 'the dualistic approach whereby morphological evidence is used to *recognize* species which are *defined* in genetic terms'.

Comparability of species and higher taxa

There have been many attempts to define a species, none totally successful. This difficulty has led to the cynical definition of a species as a group of individuals sufficiently distinct from other groups to be considered by taxonomists to merit specific rank.[354] The crux of the question does, of course, lie in the term 'sufficiently distinct', since, from what has been said above, there is no magic formula to decide the issue. Most taxonomists use one or more of four main criteria.

1.   The individuals should bear a close resemblance to one another such that they are always readily recognizable as members of that group.

2.   There are gaps *between* the spectra of variation exhibited by related species; if there are no such gaps then there is a case for amalgamating the taxa as a single species.

3.   Each species occupies a definable geographical area (wide or narrow) and is demonstrably suited to the environmental conditions which it encounters.

4. In sexual taxa, the individuals should be capable of interbreeding with little or no loss of fertility, and there should be some reduction in the level or success (measured in terms of hybrid fertility or competitiveness) of crossing with other species.

As discussed elsewhere in this book, none of these criteria is absolute and frequently it is left to the taxonomist to apply his judgement. Often he does this by attempting to recognize as species units that are of comparable significance in whatever terms are being applied.

Since species have evolved along greatly varying evolutionary pathways, and at the present time have reached all grades of distinctness, they are certainly not strictly comparable entities. They are, in fact, 'equivalent only by designation',[200] and must therefore be regarded to a considerable degree as convenient categories to which a name can be attached. Comparison of an agamospecies of *Taraxacum* with a species of *Salix* capable of hybridizing with many other species amply illustrates the truth of this statement. It is not realistic to consider species which are well differentiated on phenetic, genetic and distributional grounds as ideal or normal, and those whose taxonomic recognition poses great difficulty as non-ideal, abnormal or atypical. It has to be accepted that both are normal situations, and that the species category is a flexible, indefinable unit of practical convenience. Camp and Gilly[60] attempted to survey the different kinds of species, i.e. the different situations which lead to taxonomists recognizing taxa as distinct species. They defined 12 categories, but the names that they coined to describe them have not been generally adopted.

Above the species level there is still less comparability between the taxa at one rank. Much has been written on the 'generic concept', the 'familial concept', etc., but in fact it amounts to little more than personal judgement aimed at producing a workable classification.[254] Most taxonomists look for discontinuities in variation which can be used to delimit the kingdoms, divisions, classes, orders, families and genera. Others attempt more consciously to use the criterion of equivalence or comparability, which, although highly subjective, is frequently very useful within a group of relatively closely related taxa.

When these two criteria are contradictory a decision has to be made as to which to employ. For example, the genus *Chamerion* (*Chamaenerion*) is sometimes recognized as separate from *Epilobium*, and sometimes not. On the grounds of the distinct discontinuity in floral morphology and crossability it appears to be best considered a separate genus, but apart from the presence of this gap in the variation it seems scarcely more different from *Epilobium* than are some other groups of the genus. Raven[344] talks of the 'generic balance' in the family Onagraceae and advocates amalgamation of *Chamerion* and *Epilobium*. Usually, however, the criterion of discontinuity is the safer one to use, even though the taxa so delimited will be of very different sizes and degrees of diversity.

The six ranks mentioned above are often looked upon as more important than the other supraspecific ranks (subclass, subfamily, tribe, section, etc.) and they are termed the 'principal ranks' in the *International Code of Botanical Nomenclature* (see Table 1.2). If two families, say, are found after all to be not really clear-cut, perhaps because of the discovery of an intermediate, they are often relegated to subfamilies or tribes of a single family. A good example is the very large family Fabaceae, which represents a combination of the Mimosaceae, Caesalpiniaceae and Papilionaceae in which the last three are recognized as subfamilies (Mimosoideae, etc.). This implies that two related taxa at the rank of subfamily or tribe need not be absolutely disinct, although two related families must be. Although such a belief is rarely found in print, and certainly is no part of the *Code*, it is adhered to by most practising taxonomists.

Quite frequently a classification is adopted for patently spurious reasons. Between two taxa there are always similarities and differences, both of which have to be considered. It is simply the case that some taxonomists (the so-called *splitters*) prefer to emphasize the differences, and others (the *lumpers*) the similarities. Thus, what are clearly two or more taxa in the view of one taxonomist, are one taxon to another. Because of the lack of any absolute yardstick, such issues are among the commonest points of disagreement in taxonomy and remain unresolved. Van Steenis[452] has discussed 'the doubtful virtue of splitting families'. There is some pragmatic justification for the view that lumping is more acceptable at the level of the principal ranks and splitting more so at the others.

In other cases the classification is one largely of convenience, perhaps having arisen by historical accident, and radical changes in it being considered unwise because of the upheaval and confusion that would be caused. Walters[463] concluded that the delimitation of many families of flowering plants would have been quite different if the early taxonomists had been based elsewhere than in north-western Europe, for example the tropics or the southern hemisphere. There appears to have been an over-emphasis on the differences between taxa which are very familiar, i.e. families which are common in Europe, or are of economic value, and on the similarities between unfamiliar taxa. Hence, if Linnaeus had lived in the tropics he might well have recognized many fewer genera of Apiaceae and many more of the related Araliaceae.

These suggestions are based on two main phenomena. Firstly, it is natural to concentrate on differences between familiar objects and on the similarities between unfamiliar ones; the belief held by many Caucasians that there is greater diversity of facial appearance among Caucasians than among other races (and the reverse belief held by other races) is an every-day illustration of this. But equally, the extent to which taxa are 'split' or 'lumped' is correlated with the number of data available. The latter can be used as an expression of the degree of progress through the four phases of taxonomic exploration delimited in Chapter 1. During such progress the number of species recognized first rises, with the discovery of more and more variation, and then declines, with the realization that many of the apparent species are linked by a complete range of intermediates. Among the European flowering plants at the present day there is little change, a relatively stable situation having been reached. In many tropical regions, for example, Africa, the number of recognized species is declining as more and more thorough revisions are published. In such new works it is not uncommon for species to be allocated a dozen or more (even over a hundred) synonyms, representing species which were once considered distinct. Perhaps an extreme example is the genus *Allophylus* (Sapindaceae), which was treated by Leenhouts[256] as one polymorphic species with 255 synonyms. There is, however, evidence to suggest that this process of lumping has already gone too far in some tropical areas (see Chapter 10). In even less known regions, for example, tropical South America, the number of species recognized is still rising, as it is in many lower groups, especially algae. In bryophytes the number of genera and species is similarly increasing, the former mainly by the splitting of older, wider ones. For example, the number of genera of mosses recognized in Britain has risen by over 50% (115 to 175) in the past 50 years.

The number of names of species of seed-plants listed in *Index Kewensis* up to and including Supplement 18 (covering the years 1753 to 1985) is about 970 000 (data supplied from Royal Botanic Gardens, Kew), showing a yearly increase of about 4 000 at present. Of these 4 000, about 2 300 are reckoned by their authors to be new species; the rest are new names or new combinations for already known species. Despite this increase, the number of distinct species of seed-plants estimated by various authorities has dropped over the past 20 or 30 years from about 330 000 to about 240 000, i.e. a rise from, on average, fewer than three synonyms per species to more than four.

The amalgamation or separation of two species occasions the change of name of only one taxon, but when genera are so changed the names of many species might be altered as a consequence. At higher levels changes of rank, etc., have no repercussions on the names of species, but there are naturally fewer occasions on which such relatively important changes seem necessary. There is undoubtedly a reluctance on the part of most taxonomists to make sweeping changes in the delimitation of genera and higher ranks, because of the undoubted advantages of having a *stable* classification. Good illustrations of this are the families Lamiaceae and Verbenaceae, and *Festuca* and several closely related genera, which in both cases show no absolute differences but which are still almost universally regarded as separate. Their amalgamation would, in each instance, create a very large 'inconvenient' taxon. Nevertheless, in all situations, one needs to weigh the desire for stability against the need for change; but if the evidence for the latter is strong and unequivocal then surely it must prevail.

There seems to be a particular dislike or suspicion among taxonomists for *monotypic* taxa (i.e. those with only one species). The general preference to include, say, a monotypic genus in a family with other genera rather than in a family of its own has led to the idea that the size of the disjunction between taxa should be inversely proportional to the number of species in the smaller taxon. In other words, monotypic families should be recognized only when they are separated from others by wide gaps (cf. Adoxaceae, p. 191), but large families can be separated by very small gaps (cf. Verbenaceae/Lamiaceae, above). There is no biological logic in this, and indeed the number of monotypic families has increased considerably in recent years with more detailed studies which have exposed hitherto unknown fundamental differences. This is equally true of angiosperms, algae and all groups between. At the generic level it has been known for over half a century that there are many monotypic taxa. Willis[480] stated that 4 853 (38.6%) of the

12 571 genera of angiosperms that he recognized were monotypic, and a further 12.9% contained only two species. Relatively few genera contain large numbers of species. Monotypic taxa are clearly natural entities and must be recognized as such taxonomically.

About 44 000 generic names of vascular plants exist; as with species, this is probably about four times the number actually worth recognizing. About 670 families of angiosperms have been described; in this case probably about two-thirds of these should be maintained.

One consequence of the shifting of the higher ranks of plants with increasing knowledge of their relationships can be that there become too many, or too few, ranks to express the observed variation. For example, the realization that each of the main groups of algae should be recognized as separate divisions in their own right has led to many of the algal divisions consisting of a single subdivision, class and subclass, i.e. four ranks where one would suffice. A similar situation exists in the pteridophytes. On the other hand the suggestion that the Chlorophyta should comprise all Embryobionta as well as the green algae leads to the problem that there are not enough ranks to express all the levels of variation that clearly exist between the Chlorophyta *sensu lato* (a subkingdom or division) and the many orders of the dicotyledons and monocotyledons; the introduction of the ranks of superclass and superorder solve this particular problem. Similarly, the discovery of the differences between Procaryota and Eucaryota, which appear to transcend the traditional boundaries between the kingdoms (plants, animals, etc.), has led to the suggestion of the new supra-kingdom rank of Dominion. Zoologists often have need to adopt super-ranks (e.g. superfamily) to express adequately the variation which they encounter, and the same problem is often met by cladists who attempt to provide every node of a cladogram with a taxonomic rank (see pp. 57–58).

In situations such as that in the algae, the surplus ranks which contain all the taxa that the next highest rank contains are often known as *empty ranks*. The existence of empty ranks often poses major taxonomic problems, which may be illustrated by reference to the very distinctive plant, *Adoxa moschatellina*. The genus *Adoxa* until 1981 contained only one species, which has never been placed in any other genus. *Adoxa* itself has been variously placed in the Saxifragaceae and Caprifoliaceae (which are not considered very closely related), but its differences are so great that it is now universally placed in its own family, the Adoxaceae. There still remains the problem however, of where to place that family. It is clearly a dicotyledon (subclass Dicotyledonidae), but does it constitute a distinct order (or suborder within an order), or is it separate only as far as the level of family? In other words, how far up the hierarchy does one go before *Adoxa* is amalgamated with other taxa? In such cases *comparability* is the only means of decision-making. According to Cronquist, the Adoxaceae belong to the order Dipsacales (along with the Caprifoliaceae and other families), but according to Dahlgren to the Cornales (along with the Sambucaceae, a segregate of the Caprifoliaceae, but not with the bulk of the latter). In the early 1980s four further species of Adoxaceae were described, and variously placed in one or two extra genera, but this complication does not alter the general picture drawn by this example. The same problems exist with many other isolated taxa, for example the dicotyledon genera *Paeonia* and *Hippuris*, *Ginkgo*, the liverwort *Takakia*, and the Characeae.

The need to resort to the subjective assessment of comparability or equivalence has therefore led to most of the taxonomic ranks from order to species having 'come to have' a finite meaning in the minds of most taxonomists—a meaning which, however, defies definition. The application of this concept of equivalence, along with a desire to express phylogenetic relationships as far as they are understood and with the use of many newly investigated characters, have been the major causes of the abandonment (splitting) of such taxa as the Pteridophyta, Algae and Gymnospermae.

Use of infraspecific ranks

The species is commonly regarded as the base-point in taxonomy. This dates from the days of the early botanists, who saw the species as the finite entity produced by the Almighty at the original creation, but it is reinforced by our ability to recognize species as entities in the field and in the herbarium, and by our observations on the genetic structure of populations. Hence the species is the lowest rank which it is essential to recognize for general taxonomic purposes. It is not, however, the lowest rank which it is desirable and useful to recognize.

The use of infraspecific catagories has varied greatly over the years, and there is still little uniformity in their adoption today. Linnaeus utilized only one infraspecific taxon, the variety (distinguished by the letters of the Greek alphabet), and this only sparingly. During the nineteenth century the variety became far more widely used. Towards the end of the nineteenth and in the beginning of the twentieth century other ranks, such as subspecies, subvariety, form, etc. (and non-English words often with no exact English equivalent), were introduced. The chaotic situation which arose from the varied interpretation of those ranks led to two divergent developments. Firstly, various authors attempted to define in reasonably precise terms a limited number of infraspecific ranks, in an attempt to stabilize the situation. By far the most successful of these was Du Rietz,[106] and his definitions are still often quoted today as the basis for current usage (see pp. 193–194). Secondly, other botanists completely or largely abandoned the use of infraspecific ranks, on the ground that the confusion was caused by the lack of any real boundaries between the different ranks and in view of the unlikelihood of this problem becoming resolved.

The extreme proponent of the latter approach was V. L. Komarov in Leningrad, whose views were crystallized in the 30-volume *Flora U.R.S.S.* (1934–1964). In this Flora no infraspecific taxa are recognized, but the number of species is swelled enormously by the upgrading of any distinctive infraspecific variants to the level of species. Since in *Flora U.R.S.S.* groups of closely related species are frequently recognized as series, the latter are often equivalent to the species of western authors and many of the species to subspecies or varieties. This approach has some features to its credit, in particular the lack of a need for decision-making among the problematical infraspecific ranks and the attention which is focused upon a taxon when it is accorded the rank of species. But the rigidity by which a taxon must be recognized as a species or not at all has attracted a great deal of criticism. It has now been abandoned even in Russia; in *Flora Partis Europaeae U.R.S.S.* (1974→) subspecies are recognized extensively and the rank series appears to have disappeared.

Nevertheless there are many contemporary authors who, recognizing the lack of progress in reaching a conformity of treatment of infraspecific ranks, propose that taxonomists should compromise by recognizing only one such level, the subspecies.[341] This method of treatment is at present the one adopted by most zoologists, and the subspecies is the only infraspecific rank recognized by the zoological and bacteriological *Codes*. It is also encouraged by various modern trends in plant taxonomy, in particular the realization that an inventory of the plant species of the world is a high priority, before many of them are lost in the face of advancing civilization (see Chapter 10). Such an inventory is best obtained by concentrating on the species level, with at best only a single infraspecific rank reserved for major, usually geographical, variation. Thus *Flora Europaea* (1964–1980) dealt with only subspecies below the species level, and, although such authoritative works utilize a system which was specifically designed for their own purposes, their pattern tends to be taken up by others even when more appropriate systems would suit their particular circumstances. Infraspecific taxonomy is also bedevilled by the lack of an index of names published, for *Index Kewensis* has hitherto been concerned only with the species level and above. (Supplement 16, which appeared in 1980, has commenced the listing of all new infraspecific taxa as well, but there are no plans to cover the backlog.) A concentration on the species level and above is, therefore, the 'easy way out', and accounts for the loss of emphasis on critical, infraspecific variation.[406]

There is, however, a need to express infraspecific variation in taxonomic terms, for the provision of a name attracts attention to a taxon.[410] Moreover, much infraspecific variation is very important, and the lack of means to express it often leads to over-splitting of species. It can be argued, for example, that many subspecies in *Flora Europaea* were recognized as such only because, if they had been retained as varieties, no mention of them would have been possible, just as there were too many species recognized in *Flora U.R.S.S.* The *Festuca rubra* and *F. ovina* groups illustrate very well the tendency to split species. *Festuca rubra*, as recognized by E. Hackel in his Monograph of 1882, is represented by 21 separate species in *Flora Europaea* (1980), and *F. ovina* by 91. Several taxa which were treated as only subvarieties by Hackel are recognized as species in the later work.

The *International Code of Botanical Nomenclature* recognizes five infraspecific ranks: subspecies, variety, subvariety, form and subform. Although this has, in the past, been considered insufficient by many workers, and well over 100 different infraspecific ranks have been proposed at one time or another, nowadays only three of these ranks are commonly employed, subvariety and subform being largely abandoned. These three ranks were defined by Du Rietz:[106]

Subspecies   'a population of several biotypes forming a more or less distinct regional facies of a species'. It is thus a geographical race, ecotype, topodeme or genoecodeme.

Variety 'a population of one or several biotypes, forming more or less distinct local facies of a species'. It is thus a local or ecological race, an ecotype or genoecodeme of a lower order, or an ecodeme.

Form 'a population of one or several biotypes occurring sporadically in a species population in one or several distinct characters'. It is thus a genodeme or relatively minor genetic variant occurring mixed with other such distinct variants.

The fact that these three definitions have been retained by the majority of taxonomists for half a century indicates that they must have served (and still serve) a useful function. Many species exhibit such levels of variation, or at least one or two of them, and the three ranks can be used to express them satisfactorily. In many Floras they are used without comment, sometimes a species being divided into one or the other categories, sometimes into all three in hierarchical order, their connotation in the sense of Du Rietz being taken as understood. Camp and Gilly[60] preferred to recognize only two of these ranks—the subspecies for taxa which form distinct populations and the forma for those which do not. They considered that 'at best the variety is a category of indecision', and should be used to denote an infraspecific taxon of unknown significance.

There are, however, a number of other patterns of species variation which do not coincide with any of Du Rietz's categories. Semi-cryptic species, which are distinguishable only by close scrutiny and often only by reference to one particular organ, so that identification is at best difficult and sometimes impossible, are the commonest of these critical (i.e. problematical) taxa. Frequently they represent different levels of ploidy or are genetic lines of inbreeding taxa, so that many botanists wish to recognize them taxonomically, although they do not seem to merit full specific rank. Nowadays such taxa are most commonly recognized at the subspecific level, although sometimes as full species or as varieties. In addition, the subspecific rank today embraces ecological races, physiological races, seasonal variants and various other relatively minor morphs which taxonomists wish to name. Some botanists even place agamospecies at the rank of subspecies, although these are usually considered full species. In the current climate of concentration on the subspecies at the expense of the lower ranks the former 'has become the dumping ground for many sorts of situations, much as the variety was 100 years ago'.[406] This has led to a debasement of the subspecies concept, i.e. the loss of its identity with the geographical race and therefore with the zoological subspecies.

There are many other problems which have led to the present decline of infraspecific classification. Geographical races, recognized as subspecies, often themselves vary geographically, so that one is faced with geographical races of a lower order; but one cannot recognize subspecies of a lower order. Thus *Calystegia sepium* has several well defined geographical races which are known as subsp. *sepium*, subsp. *roseata*, etc. But many taxonomists prefer to relegate a related species, *C. silvatica*, to subspecific rank under *C. sepium*, in which case subsp. *roseata*, etc., have in turn to be relegated to varieties under subsp. *sepium*, although in terms of Du Rietz's definitions the latter are true subspecies. In other cases, where races or ecotypes are very localized, for example, on one mountain or one stretch of coast, there is genuinely a transition from subspecies to varieties which is difficult to express taxonomically.

The problem of semi-cryptic species, whatever their basis, can be solved by the use of some sort of two-tier species concept (although of course only one of these tiers could be termed a species for nomenclatural purposes). The most widely used of these systems is the species aggregate, whereby the species in the broader (higher) sense is known as an **aggregate** and the lower taxa as **segregates**.[200,407] Hence, one can talk of the '*Calystegia sepium* aggregate', of which *C. sepium* and *C. silvatica* are segregates. To distinguish the two senses in which *C. sepium* can be used one can talk of *C. sepium sensu lato* (in the broad sense) and *C. sepium sensu stricto* (in the narrow sense). This system has the advantage that it is informal, so does not pose problems of nomenclatural procedure. A group of closely related taxa at and/or just below the species level is often referred to as a **species complex**, or loosely as a 'species group'.

But most problems of infraspecific taxonomy are insoluble within the limits of the procedure laid down in the *Code*, because the ranks are hierarchical and much of the natural variation is patently not so. Sometimes the pattern of variation is of a different nature in different areas, so that a suitable classification in one area is quite unsuitable in another. It is not nomenclaturally possible to have two parallel systems of names. In other cases the variation is reticulate rather than hierarchical, and cannot be expressed satisfactorily at all. This problem is most marked with cultivated plants,[319] often because classifiers have concentrated on variation in the part of the plant used by man, at the expense of the other parts.[318] For example peas (*Pisum sativum*) can be classified by pod and seed characters, growth habit, flower colour and earliness. Since there are no means of deciding which of these is most important (and therefore delimits the higher taxonomic ranks) and, in any case, plants occur with all combinations of variation of the different groups of characters, a meaningful hierarchical classification cannot be made.

To a considerable degree, the desires to express taxonomically the *grades* as well as the different *sorts* of variation or evolution are irreconcilable, at least within the constraints of the present *Code*. What is needed is a new system of infraspecific nomenclature which is much more flexible, yet at the same time precise and meaningful.[410] It would probably also need to remove the present constraints imposed by the *Code* in connection with the law of priority. But there is no doubt that there *is* a need for a system of infraspecific classification; there is no room for complacency ('the present system seems to work reasonably well') or defeatism ('it is impossible to devise a really satisfactory system'). 'The name of a plant is the key to its literature'[451]— without a name or with a wrong name we cannot gain access to what is known about a taxon.

It should be noted here that the infraspecific classification and nomenclature of cultivated plants and of interspecific hybrids are different from those of wild species, and are treated more fully in the next chapter.

9
Ways and means

Many scientists are under the impression that taxonomists do not make heavy financial demands on Institutional budgets. In the sense that they rarely require elaborate equipment *in addition* to that used by specialists in other biological fields, and that their use of electron microscopes and the like is normally only intermittent and can be shared with others, this is justified. But support for taxonomic work can be costly, and thus difficult to obtain, in two less direct ways: space and manpower. This is illustrated in three areas: the garden, the herbarium and the library. Each of these is surveyed briefly, followed by a closer consideration of the different kinds of taxonomic literature.

The experimental garden

There is a need for growing facilities in the garden as well as in controlled glasshouse conditions. Even in the tropics indoor growing space (in 'screen-houses') is necessary for all but the most sun- and drought-tolerant species. In Copenhagen a special building ('arctic greenhouse') is used for growing arctic species,[41] and special constructions are also needed for bryophytes and pteridophytes. As well as providing artificial (e.g. frost-free, long day-length) environments, glasshouses provide draught- and insect-free areas in which hybridization experiments can be carried out in all weathers.

Apart from its obvious aesthetic appeal, a botanic garden can be said to have four main uses, which are reflected in the kind of plants grown:

1 Taxonomic research projects The garden is the means of gathering together in one place a wide range of species which can be used for anatomical, cytological, chemical or breeding work, etc., at any time. It is also needed for growing plants under controlled conditions to check on ecophenetic variation, and for growing the progeny of artificial hybridizations or of wild hybrids.

2 On-site teaching Collections of plants are often displayed in such a way (e.g. by families or habitats) that they can be used for self-instruction or for demonstration purposes.

3 Provision of material Plants are often grown specially for use outside the garden—either for teaching or for the research work of plant physiologists, chemists, zoologists, etc.

4 Conservation The culture of rare or particularly interesting plants for study in relation to conservation and the setting up of gene banks is increasing. Botanic gardens are now seen as an important facility in the conservation of genetic diversity.

Most well-organized botanic gardens operate an informal seed-exchange scheme, in which at least 500 institutions all over the world participate. Annual lists of available species are exchanged between the botanic gardens, which are able to order, free of charge, small samples of seeds of whatever species they require. Many gardens annually distribute thousands of seed samples each in this way. In recent years there has been a well-marked trend to offer seed of known wild origin in preference to unlocalized botanic garden material.

A documentation of 798 of the world's most important botanic gardens is given in *International Directory of Botanical Gardens*, 4th edition.[192]

The herbarium

A herbarium is a collection of dried, pressed plants mounted on sheets bearing a detailed data label and stored in strong cupboards in systematic sequence (Fig. 9.1). Herbaria may cover all plant groups and all geographical regions, or may be variously restricted in scope; their specimens vary from a few hundred to several millions. They may be owned by national or local governments, universities, private institutions and research organizations,

Fig. 9.1 A modern herbarium with metal, dust-proof cabinets and well-lit working surfaces.

scientific societies or individuals. Over 1 700 of the world's most important herbaria are listed in *Index Herbariorum* Part I (*Regnum Vegetabile*, Vol. 106, 1981), which gives useful data, for example size, staff and circumscription, on each. Part II of the same publication is a list of the most important plant collectors with the present location(s) of their specimens. More detailed information on herbaria in the British Isles is given in *British and Irish Herbaria* (1984), a publication of the Botanical Society of the British Isles. A guide to the world's timber collections is provided by *Index Xylariorum* (*Regnum Vegetabile*, Vol. 49, 1967).

The ten largest herbaria in the world are shown in Table 9.1. The information contained in *Index Herbariorum* has made the particular importance and contents of individual herbaria widely known. Virtually all scientifically valuable herbaria operate a loan scheme whereby specimens may be borrowed for short-term study, and many of them participate in the informal exchange of duplicate specimens. Material is also obtained by collection (often on special expeditions abroad), by purchase, and from bequests.

Table 9.1 The ten largest herbaria in the world, i.e. all those with at least 4 million specimens. Data taken from *Index Herbariorum*, edition 7 (1981).

Herbarium (and abbreviation)	No. of specimens
Muséum National d'Histoire Naturelle, Paris (P)	10 500 000
Royal Botanic Gardens, Kew (K)	>5 000 000
Komarov Botanical Institute, Leningrad (LE)	>5 000 000
Conservatoire et Jardin Botaniques, Geneva (G)	5 000 000
New York Botanical Garden, New York	4 300 000
Harvard University, Cambridge, USA (A+FH+GH)	4 250 000
U.S. National Herbarium, Washington DC (US)	4 110 000
British Museum (Natural History), London (BM)	4 000 000
Institut de Botanique, Montpellier (MPU)	4 000 000
Naturhistoriska Riksmuseet, Stockholm (S)	4 000 000

Well preserved herbarium material, providing it is adequately documented, can yield considerable taxonomic information and does not deteriorate much with age, even over hundreds of years, so long as measures are taken against insect and fungal attack and the specimens, which become increasingly fragile, are handled carefully. Contrary to popular belief, herbarium specimens can provide almost as many morphological and anatomical data as living ones, and can often be used for chemical analyses as well. Aspects of the specimen which cannot be preserved, for example height of tall plants, flower colour, habitat details, and chromosome number, should be noted on the data label, which thus becomes part of the information-content of the specimen (Fig. 9.2). Most herbarium specimens collected before this century, and many during it, are very inadequately documented and inevitably of vastly lower value.

A herbarium can offer four main services:

1 The identification of specimens. The wide range of material in a herbarium permits a direct comparative method of identification which is possible by no other means.

2 The basis for research and the preparation of Floras and monographs. Again, it is the existence of a wide range of taxa and a large sample of examples of each taxon under one roof which renders the herbarium irreplaceable in taxonomic research.

3 Teaching. Ideally this is carried out jointly in the herbarium, in the botanic garden, and in the field.

4 Preservation of voucher specimens. The most important of these are type specimens (q.v.). In addition, specimens which have been examined for chromosome number, analysed for chemical constituents, recorded for a particular locality in a Flora, or drawn to represent a species in a publication, among other fields of research, should be deposited in a herbarium, so that future workers can check the authenticity of the material investigated.

UNIVERSITY OF LEICESTER

FLORA OF SICILY

<u>Vulpia</u> <u>sicula</u> (C. Presl) Link

Open, grassy areas in mixed

deciduous/evergreen woodland.

Bosco della Ficuzza, below Rocca

Busambra, south of Palermo, c. 800m.

29-5-1972.   2<u>n</u> = 14 (code V379).

C.A. Stace & R. Cotton No. 466.

Fig. 9.2 A data label from a herbarium sheet, in this case a specimen of *Vulpia sicula* collected from Sicily in 1972 and whose chromosome number has been determined.

Because of these many important uses, the herbarium is usually the focal point of taxonomic research. Herbaria often possess subsidiary collections of material which are not amenable to preservation in the usual way, i.e. mounted on a sheet of paper. Among these are spirit-preserved specimens, especially of fleshy or flimsy plants or organs; boxes or trays of bulky dried material, for example timber, bark, large fruits; drawings, paintings and photographs; and microscope slides of anatomical and cytological preparations, or of whole organisms in the case of algae and small bryophytes. Special sheet-mounted collections may often be preserved separately from the main collection, for example diseased, galled and teratological specimens, mass population samples, and valuable historical collections.

The ability to store pressed specimens for a long period allows a succession of scientists to examine the same specimen and precisely determine the nature of the material of previous researchers. Some important specimens, such as many of those of Linnaeus in the Linnaean Herbarium in London, have been examined by generations of eminent taxonomists over the past two hundred years. When experts have opinions which differ from those of their predecessors they usually record the fact by annotating the sheets; nowadays this is done on small determination labels which are attached to the sheet. The opinions of various, often non-contemporary, experts are thus recorded in one place.

Apart from the obvious problems which arise in the curation of large herbaria, one of the greatest difficulties is the extraction of the data contained in a herbarium. Where the data required are related to a taxon the problems are few, but where the data are otherwise related, for example to regions or to collectors, they are often practically unobtainable. The storage of the data contained on a herbarium sheet in a computerized, retrievable form is therefore an enormously valuable potential development, although the vast manpower needed to accumulate such a data-bank is often not available. Thus, although many herbaria have data-banks of recently acquired material, and incorporate the data from current acquisitions, no major herbarium has so far been able to record similarly the vast backlog of older material.

The library

No scientific subject relies as heavily on an accumulation of literature (as opposed to current literature) as taxonomy. The library is therefore as important in much taxonomic work as the herbarium, and a good knowledge of the available literature is vital to the practising taxonomist; indeed, it is a good method of judging taxonomic expertise. It is largely the lack of extensive libraries which limits the kinds of taxonomic work which can be carried out in the developing countries. Although few institutions can possess the whole range of taxonomic literature, in the developed countries many are well provided for in their particular interests and many more are within reach of the major national libraries.

The chief kinds of taxonomic publications can be roughly divided into four categories: monographs; Floras; research reports; and supporting literature.

Monographs

A *monograph* of a group of plants, for example a genus or a family, is a comprehensive account of all the taxonomic data relating to that group, involving an integration of the previously existent information with the results of the monographer's own research. Usually it takes the form of a series of introductory chapters, presenting and discussing the results of research, followed by a descriptive systematic treatment of the species and infraspecific taxa within the group. Usually the geographical scope is world-wide by definition, since it is impossible to discuss a taxon in detail without taking into consideration *all* of the lesser taxa which it contains. There are a few good examples of monographs of important genera amply illustrating the detailed and painstaking work, often a life-time's research activities, which is needed for this sort of publication. Among these may be mentioned the treatments of *Crepis*,[13] *Nicotiana*,[160] *Datura*,[39] *Pinus*,[294] *Lemna*[489] and *Avena*.[19] Despite the state of perfection implied by the term monograph there is considerable variation, even among the above examples, in the degree of comprehensiveness that they exhibit, and there are many other works which are scarcely less complete and deserving the same title. The accounts of *Oenothera*,[301] *Gossypium*,[219] *Geum*,[148] Australasian *Epilobium*[348] and the Resedaceae[2] are good examples. However, it has to be admitted that no really complete taxonomic accounts exist and, even if one could be written, it would very quickly become incomplete owing to the accumulation of the results of further research. Indeed, what was considered a complete monograph of a taxon twenty years ago would rarely pass for one today.

Naturally, the narrower the scope of the work the more complete it can be. The British Ecological Society's *Biological Flora of the British Isles* is a series of monographs of species chiefly as they occur in the British Isles; since its commencement in 1941 just under 200 species have been covered, mainly from an ecological point of view. Similar series are being produced in Israel, Canada and elsewhere. *Flora Neotropica* is a monographic treatment of the plants of the American tropics which commenced publication in 1968 and is designed to cover all groups of plants and fungi. It is clear that, even at many times the current rate of production of monographs, the flora of the world will not be covered in the foreseeable future. For anything approaching a complete taxonomic coverage a less detailed treatment is necessary.

A *revision* is such a less comprehensive monograph, and there are many thousands of excellent examples. Revisions usually incorporate far less introductory material, or even none at all, and the systematic treatment is also less complete. A monograph would include a complete synonymy, an exhaustive description, and a detailed listing of ecological, geographical, cytological, chemical, anatomical and other data. A revision also ideally incorporates a complete synonymy, but the descriptions are shorter, often being confined to the important *distinguishing* characters (i.e. a *diagnosis* rather than a description), and the supplementary data are summarized more tersely. The geographical scope of a revision is often also world-wide, although it may be restricted, as in monographs. Illustrations, mostly in the form of line-drawings, are frequently included in both revisions and monographs.

Below the level of treatment offered in a revision is the *conspectus*. This is effectively an outline of a revision, listing all the taxa, often with some or all of their synonyms, sometimes with short diagnoses but commonly without them, and frequently with a brief mention of the geographical range of each taxon. Probably the best example of a conspectus is Linnaeus' *Species Plantarum*.

Finally, at the bottom end of the scale, the *synopsis* is a list of taxa with very abbreviated diagnostic statements distinguishing them from one another, but with no other information. Frequently synopses appear in the front of revisions or Floras as a summary of contents.

Obviously there are no strict divisions in the spectrum from monograph to synopsis, and the terminology is applied subjectively. Nowadays the term monograph is often used to apply to revisions, and sometimes conspectuses, to distinguish them from Floras (q.v.). The decision as to whether a monograph, revision or conspectus should be prepared depends upon the objective and upon the time, manpower and other resources available; different methods will best suit different situations. In the rest of this chapter the term monograph will be used to apply to all the above types of publication.

Floras

A *Flora* is a taxonomic treatment of the plants of a defined geographical area, i.e. an account of the flora of a particular region. In practice, Floras are frequently restricted in taxonomic terms, for example to vascular plants or to algae, so that many publications lie on the borderline between monographs and Floras. The emphasis in these two is, however, different; in the one case it is on a taxon (it is *monographic*), in the other on a flora (it is *floristic*). The introductory chapters in a Flora, therefore, describe features of the region concerned rather than the taxonomic characters of the plants which occur in it, and one of the main functions of a Flora is to provide a means of plant identification, not a distillation of all the taxonomic information concerning the species present.

As in the case of monographs, there is every gradation from the complete monographic to the outline synoptic level of treatment, and the most suitable level needs to be decided according to the particular circumstances. The smaller the number of taxa to be considered, the more detailed can be the treatment. Examples of fairly exhaustive Floras are *Illustrierte Flora von Mitteleuropa* (1906–1931), *Flora Malesiana* (1948→), *Flora Republicii Populare Române* (1952–1976), *Flore de l'Afrique du Nord* (1952→), *The Hepaticae and Anthocerotae of North America* (1966→) and *Flora Neotropica* (1968→). Floras written in such a monographic style are often known as *critical Floras*.

For floras with large numbers of species (many thousands), a less exhaustive style is required. *Vascular Plants of the Pacific Northwest* (1955–1969), *Flora Europaea* (1964–1980), *Flora of Turkey* (1965–1985) and *Flora U.R.S.S.* (1934–1964) are examples of Floras which have been completed within an acceptable length of time and about which, with hindsight, one can say that the right compromise was attempted by the organizers. Yet more are nearing completion and will soon be in the same category.

Many hundreds of Floras have been written, with varying degrees of exhaustiveness, authoritativeness and completion, and many of them in substantially different editions. A guide to the available standard Floras of all parts of the world is provided by Frodin.[141] The botanically well-explored areas of the North Temperate zone, notably Europe, are well covered by modern Floras, although there are some important exceptions, for example Spain and Greece. In tropical areas and in Australia, southern Africa and South America the coverage is far less complete, although the situation in these areas is improving.

In the British Isles so-called *local Floras* are a special and almost unique feature. These works usually do not provide any taxonomic material (descriptions, keys, etc.), but set out the distributional data concerning each species. They are mostly compiled on a narrow regional basis (usually a county) and some contain much valuable ecological, biosystematic and ethnobotanical information. They are obviously a feasible proposition only in areas well covered by a standard, national Flora which permits ready identification of the species.

Research reports

The results of research are often incorporated direct into new monographs or Floras, but usually they are initially presented separately in scientific papers. The number of periodicals containing information on plant taxonomy runs into several thousands, and leading taxonomic establishments must provide much money for their purchase and storage.

The problem of attempting to scour this vast literature for relevant information is daunting. However, there exist a number of *abstracting journals*, some listing titles only, but others providing informative abstracts of the contents. The main one of these is *The Kew Record of Taxonomic Literature*, commenced for the year 1971, which purports to cover all articles relevant to the taxonomy of the world's vascular plants. The huge task of compiling such a publication means that the most recent volume available is usually some years behind the present. More contemporary, but far less complete, coverage is provided by more general abstracting journals such as *Biological Abstracts* and *Current Advances in Plant Science*. Specialized fields are covered by a multitude of publications such as *B.S.B.I. Abstracts* (taxonomy of vascular plants of the British Isles) and *World Reports on Palaeobotany*.

Surveys of current bryological articles are provided in *Journal of Bryology*, *The Bryologist* and *Revue Bryologique et Lichénologique*.

Supporting literature

Taxonomists are fortunate in having at their disposal a large number of highly informative indices, catalogues, glossaries and similar compilations. Many of these are produced by the *International Association for Plant Taxonomy* in their occasional serial *Regnum Vegetabile* and in their journal *Taxon*. Several items under the former title have already been mentioned. Some of the most important of the remainder, and others not in that series, are listed below.

1 Taxonomic Literature There have been many incomplete attempts to catalogue the important standard taxonomic literature. All such publications are either very old or restricted in coverage (or both), but we now have a relatively exhaustive and authoritative series of *Regnum Vegetabile* entitled *Taxonomic Literature* (1976–1988). This series of seven volumes is an expanded version of an earlier catalogue and has become known as TL2. It covers full bibliographical details (including precise dates of publication) as well as biographical data on the authors.

2 Index Londinensis This index was originally issued in six volumes (1929–1931) by O. Stapf, and a two-volume supplement was produced in 1941. It lists published illustrations of vascular plants from 1753 to 1935 and, although now greatly out of date, is still a very valuable source of references.

It has been partially updated by *Flowering Plant Index of Illustrations and Information*, compiled by R. T. Isaacson (2 volumes, 1979), which lists post-1935 coloured illustrations only.

3   Index Holmensis   This is a serial publication, commenced in 1969 and not yet completed, which is a world bibliography of distribution maps of vascular plants. It is compiled and published in Sweden. Its title has latterly been changed to *Index Holmiensis*.

4   Indices of Plant Names   As stated earlier, nearly a million names of seed-plants have been published, in addition to those of many lower plants. The following are the main catalogues of these names; in each catalogue the author of the name and the place of publication are given, and sometimes other details.

(*i*)   *Index Kewensis*   The original volumes were produced at Kew in 1895 with the aid of a bequeathal from Charles Darwin, and covered the gymnosperm and angiosperm species described between 1753 and 1885. Subsequently, Supplements have been produced at Kew, bringing the bibliography up to date (and inserting earlier omissions). Currently the Supplements are produced every five years and, to date, 18 have appeared, covering up to 1985. In future, annual supplements are planned under the title *Kew Index* and listing pteridophytes as well as spermatophytes; the first (for 1986) appeared in 1987. The original volumes, and the first three Supplements, distinguished between species which were considered taxonomically acceptable and those which were believed to be synonyms of others, but many errors were introduced by this subjective judgement and the attempt was wisely abandoned after Supplement 3 (1908). From Supplement 16 (covering 1971 onwards) the listing of all infraspecific taxa as well as species was commenced, an addition made possible by their compilation in *The Kew Record of Taxonomic Literature* (see p. 203).

(*ii*)   *Index Nominum Genericorum* (ING)   A listing of all the generic names of plants of all groups, both fossil and recent, was published in 1979 under the above title (3 volumes, *Regnum Vegetabile*, Vols 100–102). It took 25 years to complete, and covers all generic names from 1753 to 1975. As well as giving bibliographic and nomenclatural details, type species are indicated; there are about 63 500 entries. The first Supplement (*Regnum Vegetabile*, Vol. 113, 1986) lists 2 500 extra generic names and gives many corrections to the previous ones.

(*iii*)   *Other Catalogues of Flowering Plants*   Many other more specialized and variously restricted catalogues of flowering plants exist, for example *Index to Grass Species* (1962), *Repertorium Plantarum Succulentarum* (1951→) and *A Dictionary of the Flowering Plants and Ferns*, 8th edition (1973) by H. K. A. Shaw, which lists every family name published since 1789 and every genus since 1753, and assigns all the latter to their family. The *Gray Herbarium Card Index* (1894→) lists all new names (including infraspecific ones) of vascular plants in the New World, starting in 1886 (1754 for infraspecific names).

(iv) *Pteridophytes* *Index Filicum* (original in 1906 covering names up to 1905, five Supplements since covering names up to 1975) was commenced by the Danish botanist C. F. A. Christensen. A Supplement covering names from 1976 to 1985 is in preparation. Supplement 5 (1961 onwards) covers all pteridophytes, but earlier parts treated ferns only. Earlier Lycopodiales are covered in *Index Lycopodiorum*,[151] Isoetales in *Index Isoetales*,[267] Psilotales in *Index Psilotales*,[268] Selaginellales in *Index Selaginellarum*[269] and Equisetales in *Index to Equisetophyta*.[270]

(v) *Bryophytes* *Index Muscorum* consists of five volumes of *Regnum Vegetabile* (1959–1969) indexing all species and infraspecific taxa of mosses published up to the end of 1962. A supplementary volume covering the period up to the end of 1973 is in preparation, and from that date biennial Supplements are being produced; the first appeared in *Taxon*, Vol. 26 (1977). *Index Hepaticarum* (including Anthocerotales) was commenced in 1962 by C. E. B. Bonner, and to date about half the liverworts have been covered. The Index covers the period up to the end of 1973, and since then a series of biennial Supplements, like those to *Index Muscorum*, is being produced; the first appeared in *Taxon*, Vol. 27 (1978).

(vi) *Algae* There are no complete indices of algal species, although the diatoms have been covered in *Catalogue of the Fossil and Recent Genera and Species of Diatoms and their Synonyms* (in 7 parts, 1967–1978). A catalogue of the names of classes and families of living algae, including full synonymy, appeared as *Regnum Vegetabile*, Vol. 103 (1980); 24 classes were recognized.

(vii) *Fossils* Species of most groups of plant fossils have been indexed in various volumes of *Fossilium Catalogus*, published in The Netherlands; see also under Algae (above) and *Index Nominum Genericorum* (p. 204).

In addition to all the above there is a huge body of reference works which are of great assistance in taxonomic work. These include atlases, biographies, bibliographies, dictionaries, guides to nomenclatural and curatorial practice, and glossaries. Of special value among the last is *Botanical Latin* by W. T. Stearn.[417] Finally, a draft list of authors of all plant names[367] appeared in 1980. It gives the full name of the author, his life dates, and a standard abbreviation of his name to be used in nomenclatural citations. A more internationally acceptable version of this is being prepared.

Diagnostic keys

Both monographs and Floras usually incorporate a ***diagnostic key***, by the use of which the reader may identify an unknown plant with one included in the work concerned. A more descriptive, though seldom-used, term for such a device is ***determinator***. Keys do not offer descriptions of the plants concerned, but state only the essential diagnostic characters by means of which the taxa can be identified. Ideally they use the most conspicuous and clear-cut characters, without special regard to those considered taxonomically the most important. For this reason the sequence of taxa is often quite artificial, and such keys are frequently termed ***artificial keys***. The synopsis, referred to on p.

201, is similarly an arrangement of taxa with a bare minimum of diagnostic characters, but in this case the taxa are placed in an order which reflects their supposed natural relationships, and the diagnostic characters used may be cryptic or difficult to determine. Nevertheless, such synopses may be written in the form of a key, when they are known as *synoptic keys*. Although they often took the place of artificial keys in nineteenth century Floras and monographs, they are usually very unsuitable for the purposes of identification. A survey of most sorts of diagnostic key in current use is given by Pankhurst.[315]

Artificial keys are of two main types: *single-access* or *sequential keys*, and *multi-access keys*.

Sequential keys

Sequential keys are nearly always written in the form of the *dichotomous keys* so familiar in botanical (and zoological) works, a form apparently first used by R. Morison in his *Plantarum Umbelliferarum Distributio Nova* (1672). The dichotomous key consists of a series of *couplets* or mutually exclusive pairs of statements, each statement (or *lead*) of a pair leading on to a further couplet. At each couplet a decision to follow one lead or the other has to be taken, so that the number of possible taxa with which the unknown specimen can be identified is successively reduced until there is only one possibility. The keys may be written in one of two ways (*bracketed* or *indented*), which are illustrated, using exactly the same data, in Fig. 9.3. The essential difference between these two lay-outs is that in the bracketed key the two leads of each couplet appear together, whereas in the indented key all the possibilities arising from the first lead are dealt with before the second lead is mentioned. Each method has its advantages (and advocates), and the choice of one over the other is mostly a matter of personal preference. However, the indented key has advantages when the key is short, as the 'pattern' of characters is clearer, but when the key is long there is much wastage of page-space and the user often has to turn pages to find the two halves of a couplet. Much has been written about the theory of constructing and using dichotomous keys, but the best lesson comes from repeated practice. For example, keys may be constructed in a relatively symmetrical (the two leads of each couplet each accounting for approximately one half of the total number of taxa) or asymmetrical manner. The *average* path-length that the user needs to travel to get down to a taxon is greater in an asymmetrical key, which tends to suggest that symmetrical keys are better. However, in practice it is often better to key out very distinctive taxa first (e.g. the banana in Fig. 9.3), for in so doing the user may be more likely to arrive at a correct identification despite having used a slightly longer average pathway. If one or more of the taxa are rather variable, they may be best dealt with by keying out different parts of their spectrum of variation in different places in the key. But if this is not done (i.e. if all taxa appear only once in the key) it is useful to remember that whether the key be asymmetrical or symmetrical, or bracketed or indented, for n taxa there are needed $n - 1$ couplets.

Many variations exist in the method of presentation of sequential keys, for example the incorporation of small drawings, or the circular key adopted in

A. Bracketed key

1. Fruit more than twice as long as broad, with a skin in
   three or more distinct segments   *Banana*
   Fruit less than twice as long as broad, with a skin not in
   segments   *2*
2. Fruit with a thick, aromatic skin   *3*
   Fruit with a thin, non-aromatic skin   *4*
3. Skin yellow; fruit with a marked protuberance at one
   end   *Lemon*
   Skin orange; fruit rounded at both ends   *Orange*
4. Fruit with a single 'stone' in the centre   *Plum*
   Fruit with several separate seeds in the centre   *5*
5. Fruit greenish-yellow, hairy   *Gooseberry*
   Fruit blackish-purple, glabrous   *Blackcurrant*

B. Indented key

1. Fruit more than twice as long as broad, with a skin in
   three or more distinct segments   *Banana*
1. Fruit less than twice as long as broad, with a skin not in
   segments
     2. Fruit with a thick, aromatic skin
       3. Skin yellow; fruit with a marked protuberance
         at one end   *Lemon*
       3. Skin orange; fruit rounded at both ends   *Orange*
     2. Fruit with a thin, non-aromatic skin
       4. Fruit with a single 'stone' in the centre   *Plum*
       4. Fruit with several separate seeds in the centre
         5. Fruit greenish-yellow, hairy   *Gooseberry*
         5. Fruit blackish-purple, glabrous   *Blackcurrant*

Fig. 9.3 Sequential dichotomous keys: **A**, bracketed key; **B**, indented key. In each case the same six fruits are keyed out using identical data.

the *Geigy Weed Tables* (1975). An excellent example of a bracketed dichotomous key is *Thonner's Analytical Key to the Families of Flowering Plants*,[150] which contains 2 117 couplets. Hutchinson's key[218] to the families of flowering plants, on the other hand, uses the indented format and the couplets are not numbered, often causing difficulty in usage.

Multi-access keys

Whereas sequential keys have a single commencing point, i.e. a fixed sequence, multi-access keys can be commenced at any position. Multi-access keys are usually produced not on pages in a book, but on separate **punched cards**; in this form they were apparently first successfully operated by A. T. J. Bianchi in 1931 and by S. H. Clarke in 1936.[64] There are two main types of punched card key. In the **edge-punched key** there is one card for each combination of attributes (i.e. usually one for each taxon), and each attribute is represented by one of the holes punched around the perimeter of the card (Fig. 9.4). If a taxon possesses a particular attribute, that position is clipped out to form an open notch instead of a circular hole. All the cards are stacked up with their corresponding holes aligned, and an attribute possessed by the specimen to be identified is chosen. A thin rod or knitting needle is then pushed through the appropriate position and the rod is lifted horizontally and gently shaken. All those cards (taxa) possessing that attribute will fall away from the stack, leaving the cards without that attribute on the rod. The latter are put aside, and the others are gathered up and the process repeated with them using further characters, until only one card falls out of the stack. In the **body-punched key** (**polyclave**) the holes are punched in rows in the main body of the card. In this case each card represents an attribute, and each taxon occupies a standard position on the card; if the taxon possesses that attribute its position is punched out. To identify a specimen, some of its attributes are listed and the appropriate cards are selected. These are aligned and taxa which possess all the attributes being tested will show a hole right through the stack when held up to the light. Cards (attributes) are added to the stack until only one hole (taxon) remains.

The advantage of multi-access keys is that the attributes to be used for identification can be chosen by the user, so that if an unusual attribute is noticed the majority of the taxa are eliminated straight away. The disadvantage is that the cards are relatively cumbersome and time-consuming and expensive to construct. Attempts to overcome this in recent years have led to written versions of multi-access keys, notably that to the Turkish genera of Apiaceae by Hedge and Lamond.[189] In this method the attributes to be used are listed and coded with a letter. Each taxon is then scored for its attributes, which are written out as an alphabetical formula. In the example chosen (Fig. 9.5) all the attributes have at least one contrasting attribute, which together are all embracing and exclusive, so that each alphabetical formula is of the same length, but there is no reason why this must be so.

Taxa with the same formula are differentiated by the addition of extra, written attributes, but these could well be incorporated in the formula, and indeed there are clearly a great many procedural alternatives in this method. The alphabetical formula of an unidentified specimen can be constructed and compared with those of the named taxa. Various more simple or visually attractive methods based upon the principle of multi-access keys have been used in more recent years, such as the **tabular key** to *Ranunculus* subgenus *Batrachium*[211] and the **lateral key** to common British Poaceae.[380]

The existence of so many sorts of key in addition to the traditional dichotomous keys indicates a disenchantment with the latter. It is certainly true that dichotomous keys are inflexible (there is a set sequence and, if a character is missing, progress is halted), and fallible (one quite minor error by the user can lead to a totally wrong answer, and mistakes often become apparent only at the end of the operation). In fact the use of keys is often one of the main difficulties experienced by beginners in taxonomy, and new and better methods of identification are clearly desirable. In this context the increasing use of various types of multi-access keys is not surprising.

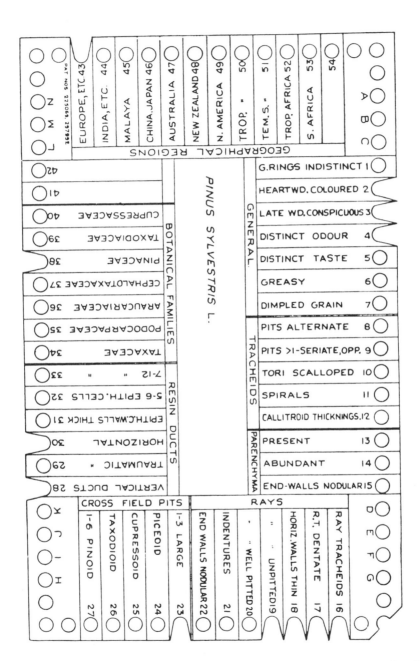

Fig. 9.4 Edge-punched card key for the identification of softwood timbers, taken from Phillips.[324] The card shown is that for *Pinus sylvestris* and has the appropriate character-holes clipped out. (© Crown Copyright; reproduced by permission of the Princes Risborough Laboratory, BRE.)

210   *Ways and means*

	Characters	Sources of Confusion
A	Flowers white, creamy white, pink, violet, red, pale blue or green.	Some white-flowered species dry bright yellow (*Daucus, Laserpitium, Echinophora*) but are still treated as A.
B	Flowers distinctly yellow.	
C	Basal or lower cauline leaves simple, entire or toothed, or reduced to petiole with odd divaricate segments.	
D	Basal or lower cauline leaves lobed, ternate or palmately divided.	
E	Basal or lower cauline leaves 1-pinnate or pinnatisect.	Transitions between E and F occur. In doubtful cases both states should be considered.
F	Basal or lower cauline leaves 2-pinnate or more.	
G	Fruit more than 3× as long as broad.	Borderline cases, of which there are very few, are classified as H.
H	Fruit less than 3× as long as broad.	

ACGJLMPQ	Plant ± junciform	9. *Rhabdosciadium*
ACHIKMPQ		75. *Heracleum*
ACHIKNOQ	Fruit margin moniliform	80. *Tordylium*
	Fruit margin smooth	81. *Ainsworthia*
ACHIKNPQ		75. *Heracleum*
ACHILMPQ	Plant spiny	5. *Eryngium*
ACHILNOQ	Leaves ± spiny	5. *Eryngium*
	Leaf margins entire	46. *Bupleurum*
	Fruit aromatic	26. *Pimpinella*
ACHILNOR		26. *Pimpinella*
ACHILNOS	Bracteoles conspicuous, persistent	4. *Actinolema*
ACHILNPQ	Plant ± spiny	5. *Eryngium*
ACHJKNPQ		75. *Heracleum*
ACHJKNPS	Orbicular lobed leaves	1. *Hydrocotyle*
ACHJLMPQ	Plant ± junciform	9. *Rhabdosciadium*
ACHJLNOQ		46. *Bupleurum*
ACHJLNOR	Stems ± absent	45. *Hohenackeria*
ACHJLNPR	Basal leaves ± oblong-linear	59. *Falcaria*
	Basal leaves ± orbicular	17. *Scaligeria*
ADGJLMPQ	Leaf margins cartilaginous	59. *Falcaria*
ADGJLNOQ		59. *Falcaria*
ADHIKMPQ		75. *Heracleum*

Fig. 9.5 Part of a written version of a multi-access key to the genera of Apiaceae in the *Flora of Turkey* (Hedge and Lamond[190]). The literal coding of the first eight attributes (A–H) is shown, followed by the first 18 formulae. Altogether 123 formulae were used to account for 97 genera.

A further important development is the use of computerized methods of identification.[104,314,315] These usually involve the principles of multi-access keys rather than sequential keys, and they may not be greatly affected if only one or two of many characters do not fit; they are therefore able to accommodate a degree of user-error or atypical material. The computer can be used in two main ways: indirectly, to generate material which is subsequently used manually, for example, polyclaves; or directly, actually to perform the process of keying itself. Pankhurst classified the latter methods as *matching*, where the computer produces a numerical value which assesses the similarity of the unknown to the known taxa, or *on-line*, where information is fed into a computer via a keyboard and the next step or answer is produced by the computer via a print-out or a visual display screen. Clearly, such computerized methods hold more hope for the future than as present-day tools (although a number are already in routine use), and will never have much direct effect on the field botanist. However, it may well be that the construction of programs for computerized identification will usefully pinpoint many of the shortcomings of the currently-available non-computerized keys.

A key in the form of a matrix (rather like a tabular key) that combines the features and (it is claimed) the advantages of single-access and multiple-access keys has recently been described.[472]

Codes of nomenclature

Few aspects of taxonomy generate more argument and misunderstanding, among both taxonomists and non-taxonomists, than nomenclatural procedure. Organisms are named according to the principles laid down in three main publications (*International Code of Zoological Nomenclature*, *International Code of Nomenclature of Bacteria*, *International Code of Botanical Nomenclature* = ICBN), and there is also *The International Committee on Taxonomy of Viruses*; comparative reviews of these are available.[228,361] In the present book discussion is kept to a minimum; the main provisions of the ICBN are merely stated in a reorganized and simplified form so that the most commonly encountered situations may be understood. The ICBN covers all plants, including fungi, slime-moulds and blue-green algae, but there is the possibility of some overlap with the other Codes in the case of organisms (such as Euglenophyta) which are variously referred to more than one of the three main groups.

The current ICBN dates largely from rules laid down at the International Botanical Congress held in Cambridge (England) in 1930, but these rules in turn were derived from various previous ones, the earliest emanating from the Paris Congress in 1867. Nowadays the ICBN is ratified and emended during the meetings of the Nomenclatural Section of each International Botanical Congress, the most recent one being the 14th, held in Berlin in 1987. The current version of the ICBN (1988) emanates from the latter. Since 1950, changes to the ICBN have become less far-reaching and are now mostly acts of clarification.

The ICBN (produced as a volume of *Regnum Vegetabile*) aims to provide an internationally acceptable code of practice on nomenclature, so that a strong element of stability is introduced. This code of practice is set out in the form of 76 main Articles (rules), with numerous Recommendations, Notes and Examples subordinate to them. The Articles are retro-active unless otherwise stated; for example the requirement for descriptions of new taxa to be in Latin dates only from 1935 in the case of bryophytes and vascular plants, and 1959 in the case of algae. The ICBN does not adjudicate upon taxonomic matters—nomenclature and taxonomy are separate procedures; the ICBN merely lays down the criteria for naming a taxon whose circumscription, position and rank have been taxonomically decided. If a taxon is taxonomically treated in one of two or more ways, there is a different nomenclaturally correct name for each.

The major provisions of the ICBN can be classified under seven major headings; much simplification has been necessary and the following omits many important details.

1　The names of taxa

(a)　Plants are classified in taxa hierarchically arranged in consecutively subordinate *ranks*. This hierarchy is set out in Table 1.2, which shows that seven of the ranks are described as the 'principal ranks' by the ICBN. Names between the ranks subkingdom and subtribe usually possess distinctive endings which denote their rank (Table 1.2).

(b)　The names of taxa at the six upper principal ranks (i.e. not species) are single words or *uninomials*, for example Ranunculaceae, *Combretum*. The names of taxa at other supraspecific ranks are *combinations* consisting of the name of the taxon at the principal rank followed by the 'name' (*epithet*) of the taxon at the lesser rank, the latter preceded by a word indicating that rank, for example *Ranunculus* subgenus *Batrachium*.

(c)　The name of a species is also a combination—a *binomial* formed from the generic name followed by the specific epithet, for example *Ranunculus acris*. Where the context is understood the generic name can be abbreviated, for example *R. acris*. The name of a taxon at an infraspecific rank is a combination of the specific name followed by the epithet of the taxon at the infraspecific rank, the latter preceded by a word indicating that rank, for example *Ranunculus ficaria* subsp. *bulbilifer*.

(d)　Supraspecific names are spelled with an initial capital letter, specific and lower epithets with a lower case initial letter. In print generic and lower names (or sometimes generic names and specific and lower epithets) are usually italicized. There are detailed rules governing the form and grammar of plant names at the various ranks.

(e)　The names of taxa, especially at and below the genus level, are often followed by the name of the author (the *authority*) who first published the name, e.g. *Ranunculus* L., or *Ranunculus tripartitus* DC. These authorities are often abbreviated[367] (in the above, L. stands for Linnaeus, DC. for de Candolle). In combinations the epithet is often not used in the same combination as was adopted by the original author; in other cases a generic name may become used as an infrageneric epithet, or *vice versa*. In such situations the name of the original authority is placed in parentheses, followed by the name of the authority using the name of the taxon in its present form, for example *Spergularia* (Pers.) J. S. & C. Presl, *Vulpia bromoides* (L.) Gray. In the latter case the name *Vulpia bromoides* used by Gray utilized the epithet *bromoides* applied to the same species by Linnaeus as *Festuca bromoides* L. The names of the authors placed after the name of the taxon are known as the **author citation**.

2 Valid publication

(a) **Legitimate names** are those which are in accordance with all the relevant rules of the ICBN and are therefore available for consideration as the acceptable name of a taxon. Names which have not been so published and are therefore not so available are **illegitimate**. A legitimate name which is the accepted name of a taxon according to the rules of the ICBN regarding the choice of names is the **correct name**.

(b) In order to be legitimate, a name must be **validly published**, i.e. (i) be effectively published (in printed matter available to the botanical public by sale, exchange or gift), (ii) obey the rules concerning the formation of names (see above), (iii) be accompanied by a diagnosis in Latin or by a reference to a previously and effectively published diagnosis (the Latin criterion does not apply to fossil plants, but fossil plants and algae additionally need a figure illustrating their diagnostic features), (iv) have its nomenclatural type indicated (see 4 below), and (v) have its intended rank clearly stated. It must also not fall foul of certain other rules, e.g. a validly published name which is predated by the same name referring to a different plant is illegitimate (see 4(b) below).

(c) Where an author uses a name which was coined by another author but not validly published by him, both authors' names appear in the author citation, the earlier preceding the later and separated from it by the word *ex*, for example *Ramatuella virens* Spruce *ex* Eichler, Eichler having adopted the name which Spruce wrote on various herbarium specimens of this species.

(d) For complete citation the place of publication should follow the author citation, for example *Ranunculus acris* L., *Species Plantarum*, p. 554 (1753). If the publication was in a work not written entirely by the author of the name, the word *in* is used, for example, *Ramatuella virens* Spruce *ex* Eichler *in* Martius, *Flora Brasiliensis*, **14(2)**: 100 (1867).

(e) In practical terms, valid publication is usually effected either by the *description* of a new taxon, or by the citation of a previously and validly published name for the same taxon. These two methods are illustrated in Fig. 9.6. The name of a newly described taxon is usually indicated by the words *sp. nov.* (species nova), *gen. nov.* (genus novum), etc.; that of a new combination, i.e. a name based on a previous name of the same taxon, by *comb. nov.* (combinatio nova); and that of a new name, i.e. a name based on a previously described taxon but using a different name, by *nom. nov.* (nomen novum). The name upon which a new combination is based is known as the **basionym**.

A   *Gaudinia hispanica* Stace & Tutin, **sp. nov.**

Gramen annuum, 6–28 cm altum, caespites parvos laxos formans. Culmi erecti rigidi graciles. Folia molliter patenterque pubescentes; vaginae valde striatae; ligulae usque ad 0.5 mm longae, hyalinae, apice truncatae laceratae; laminae 1–11 cm, primo planae, demum convolutae, usque ad 2 mm latae.

Inflorescentia 1–14 cm longa, erecta rigida gracilis spicata; rhachis supra nodos demum fragilis. Spiculae 5.5–11 mm longae, sessiles, dissitae, 2-4-florae. Glumae binae, vulgo obtusae, oblongae, glabrae vel pubescentes; gluma inferior 1.5–2.7 mm longa, superiore 2/5 usque ad 3/5 brevior, manifeste 1-3-nervia; gluma superior 3.2–7 mm longa, manifeste 4-5-nervia. Lemma ovato-oblongum, obscure 5-nervium, dorsaliter praeter ad apicem carinatum teretiusculum, muticum vel arista subapicale usque ad 1 mm longa ferens, glabrum vel pubescens, margine late hyalinum, apice acutum vel bifidum; lemma infimum 3.8–6.5 mm longum. Palea lemmati brevior, hyalina, valde bicarinata, hispida, apice acuta vel sub-bidentata. Stamina 3; antherae 2–3 mm longae, ad anthesin exsertae. Caryopsis 1.7–2.3 × 0.5 mm, liber, apice stylopodio brevi hispido instructa; hilum sub-basale punctiforme.

Holotypus: Near Ermito del Rocio, S.W. of Sevilla, Huelva, Spain, 25 May 1967, *Chater, Moore & Tutin s.n.*, (LTR). Isotypi: K, SEV. Paratypi: Entre Almonte y El Rocío, Huelva, Spain, 20 June 1969, *S. Silvestre & B. Valdés 2289/69* (LTR, SEV).

B   *Vulpia ciliata* Dumort. subsp. *ambigua* (Le Gall) Stace & Auquier, **comb. et stat. nov.**

Syn. *Festuca ambigua* Le Gall, *Fl. Morbihan:* 731–732 (1852).

Fig. 9.6 Two methods of validly publishing a new taxonomic name, reproduced from Heywood.[205] **A**, by description; **B**, by citation of a previously and validly published name for the same taxon.

3   The principle of priority

(a)   Many taxa have more than one name. Such names are known as *synonyms*, although this term is often applied only to the names other than that considered correct. These may arise because different authors have named plants in ignorance of each other's activities, because taxa once thought distinct are now considered not so, or because authors coined new names resulting from different taxonomic judgement or even personal whims, or in order to replace incorrect or illegitimate names. Choice of the correct name from among the synonyms is one of the major provisions of the ICBN. In general, only the legitimate names are in contention and, of these, the earliest is the one to be accepted. For this reason precise dates of publication (to the nearest day) are important and, in the case of older works, are often the subject of much research. Furthermore, the classification to be adopted has to be decided first, so that the full range of names in contention is known.

(b) The 'law of priority' starts with Linnaeus' *Species Plantarum* (taken as published on 1st May 1753), which therefore cannot be pre-dated. Exceptions are made for mosses except Sphagnaceae (which date from Hedwig's *Species Muscorum*, 1st January 1801) as well as some algae and all fossil plants.

(c) In choosing the earliest name for a taxon, only the names or epithets available *at that rank* can be considered, and their dates of publication are those at that rank, not necessarily the earliest publication of the name or epithet. Frequently, the earliest epithet at the appropriate rank exists only in an inapplicable combination, so a new combination has to be made.

(d) The law of priority applies only to the ranks of family and below.

(e) In some cases the usual rules of the ICBN concerning the choice of name for a taxon have to be set aside. For example, if a name has been persistently (and hence misleadingly) used in the past for the wrong taxon, it can be rejected as a *nomen ambiguum*. More frequently a very well established name for a taxon is found by later research to be incorrect (usually an obscure, earlier name being found). Loss of a familiar name is unfortunate, and the change of a generic name especially so because it could necessitate the coining of a great many new combinations for the species included under it. To avoid this, generic names found to be incorrect according to the ICBN, but whose substitution would cause much inconvenience, are able to be *conserved*, and the names which would otherwise be correct *rejected*. Such names are known as **nomina conservanda** and **nomina rejicienda** respectively, and are listed as an Appendix to the ICBN. For example, the tropical genus *Combretum* Loefl. (1758), with hundreds of species described under it, is conserved against *Grislea* L. (1753), which has hardly been used since Linnaeus' time. The ICBN also provides an Appendix of conserved family names. Both lists are added to periodically, as further cases come to light. Since 1981 it has also been possible to conserve the names of species of *major economic importance* (but not of species in general). The situation that precipitated this major change of policy was the discovery that *Triticum aestivum* was not the correct name for bread-wheat; the new rule enabled conservation of this name and the rejection of the newly discovered (but older) one.

4 The type method

(a) Names of taxa of the rank of family or below each have a **nomenclatural type**, which fixes a name to a particular taxon and determines the application of that name. The type of the name of a species or infraspecific taxon is a preserved single **type specimen**, or in some cases a drawing of one. The type of the name of a genus or of a rank between genus and species is also the type specimen of one species (the **type** *species*) in that taxon, while the type of the name of a family or of a rank between family and genus is the same as the type of the genus (*type genus*) on whose name the family is based.

(b)   Names identical in form but based on different types are termed *homonyms*; for example, *Festuca incrassata* L. is a quite different species from *Festuca incrassata* Salzm. *ex* Lois. All *later* homonyms are illegitimate.

(c)   A single type designated by the original author of a taxon is known as a *holotype*. Such a type must now be designated to effect valid publication, but this has not always been so. *Isotypes* are duplicates of the holotype, often being sent to other herbaria. *Syntypes* are two or more specimens designated by the original author when no holotype was designated. A *lectotype* is one of the syntypes subsequently chosen to act *in lieu* of a holotype. A *neotype* is a specimen designated to act *in lieu* of a holotype when no holotype or syntypes exist. The type of a new combination is automatically the type of its basionym, and the type of a new name is automatically the type of the replaced name. Type specimens or type taxa are not necessarily 'typical' in the usual sense of the word—they may even be relatively unusual examples. A type designated by the original author or selected by a later one cannot be changed on such grounds.

(d)   The name of an infraspecific taxon which includes the type specimen of that species must have the same epithet as the species, and the repeated name cannot be pre-dated and bears no author citation. Hence, of two subspecies of *Gentianella amarella*, viz. *G. amarella* (L.) Börner subsp. *amarella* and *G. amarella* (L.) Börner subsp. *hibernica* N. Pritchard, the former includes the type specimen of *G. amarella* (named *Gentiana amarella* by Linnaeus) and is known as the *typical* or *nominate* subspecies, while the latter does not. Hence, if a subspecies is described within a species which hitherto had none, the name of the typical subspecies becomes automatically fixed, and it is the non-typical subspecies which must be given a different name. Such repeated names are termed *autonyms*.

Similarly, the names of typical infrageneric taxa are fixed according to the type species, for example, *Ranunculus* subgenus *Ranunculus* and subgenus *Batrachium*, the type species (*R. acris*) belonging to the former. The names of familial and typical infrafamilial names are fixed according to the type genus, in this case with different endings, for example Ranunculaceae tribe Ranunculeae, containing the type genus *Ranunculus*. The names of taxa above the rank of family may be fixed by following the same typological procedure, but this is not mandatory, despite several proposals that it should be so.

(e)   When a species is split into two or more the part which retains the type specimen of the original species must retain the same name, and the other parts must be given different names. Similarly when families and genera are split, one of the splits (the typical one) must retain the original name.

5   Name changes

The change of name of a familiar and much cited plant can be very tiresome, but stability can be reached only by a vigorous application of the rules of the ICBN. A large number of the Articles in the ICBN are detailed expositions of the methods to be used in choosing the correct name. Name changes result from three main causes:

> Changes due to the discovery that the name being used is not correct according to the ICBN (*nomenclatural changes*);
> Changes due to taxonomic change of opinion, for example amalgamation, splitting or transference of taxa (*taxonomic changes*);
> Changes due to the discovery that a taxon has previously mistakenly been given the name of a different taxon (see 3(e) above).

Synonyms are therefore of two types: **homotypic** or **nomenclatural synonyms**, based upon the same type as the correct name; and **heterotypic** or **taxonomic synonyms**, based upon a different type from the correct name, and therefore synonyms only by taxonomic judgement.

It is not permitted to change the name of a plant for reasons of convenience, however pressing. *Phlomis italica*, *Sibthorpia africana* and *Teucrium asiaticum* are three species that have never been found outside the Balearic Islands, Spain. None of them occurs in Italy, Africa or Asia, yet their present names (based on misapprehensions of Linnaeus, their author) must stand.

6   Hybrids

Hybrids are largely governed by the same rules as species, etc., but the ICBN sets aside twelve special Articles for them alone. The most important points are:

(a)   Hybrids between species within the same genus (interspecific hybrids) are designated by a formula—the two species names separated by a multiplication sign, for example *Calystegia sepium* × *Calystegia silvatica*. In addition they may be given, if an author so wishes, a binary name, consisting of the generic name and a hybrid epithet separated by a multiplication sign, for example *Calystegia* × *lucana* (the same as the last example). The binary name is equivalent in rank to a species and is governed by the same rules.

(b)   Variants of an interspecific hybrid may be described as nothosubspecies, nothovarieties, etc., which are equivalent in rank to subspecies and varieties, for example *Salix* × *rubens* nothovar. *basfordiana*.

(c)   Intergeneric hybrids are similarly designated at the generic level by a formula, for example *Coeloglossum* × *Gymnadenia* and, if required, in addition, by a hybrid-genus name preceded by a multiplication sign, for example × *Gymnaglossum* (the same as the last example). At the species level they may be designated by either a formula, for example *Coeloglossum viride* × *Gymnadenia conopsea*, or by a binary name, for example × *Gymnaglossum jacksonii*.

7 Cultivated plants

The taxonomy of cultivated plants is often very complex, and the rules governing their names are equally so. For the most part these rules are the same as those governing wild plants, but there are differences and, because cultivated plants are of great importance and interest, a separate volume of *Regnum Vegetabile*, the *International Code of Nomenclature of Cultivated Plants* (ICNCP), with 57 Articles, has been produced by a committee of the *International Union of Biological Sciences* (latest edition 1980). Many aspects of this complex topic were discussed in two recent symposia.[429,450]

Most of the rules in the ICNCP are taken from the relevant ones in the ICBN, but below the species level there is a provision for the recognition of only one rank, the ***cultivar***, instead of the normal hierarchy. Cultivars are recognizable infraspecific entities of diverse nature, for example, clones, self-fertilized pure-lines, or cross-fertilized assemblages characterized by one or more attributes. They are written with an initial capital letter, not in italics but in single quotation marks, and follow the name to which they are subordinate either alone or preceded by 'cv'. With certain exceptions, cultivars have vernacular, not Latinized, names. Cultivars can be subordinate to species, genera, hybrids or even vernacular names, as long as there is no ambiguity. Examples of cultivars are *Rubus idaeus* 'Malling Wonder', *Viburnum* × *bodnantense* 'Dawn', *Rosa* 'Crimson Glory', and Apple 'Cox's Orange Pippin'.

Cultivars are usually produced artificially and are subject to national and international regulations protecting the rights of plant breeders. For many groups of plants (e.g. roses, orchids) there are **Registration Authorities** which keep an inventory of all cultivars in the group covered, and form the basis for ensuring breeders' rights according to the separate laws of each country.

If preferred, cultivated plants may be named by the hierarchy of infraspecific ranks laid down in the ICBN.

10
Taxonomy in the service of man

Taxonomy is often thought of as a purely academic or even superficial subject of little relevance to the urgent needs of modern man. Such a view is, however, far from the truth, for plant taxonomists have an important role to play in man's betterment, a role emerging as a key one now that many of our traditional natural resources are in short supply or running out. This chapter presents evidence for such a claim.

Rich and poor floras

The vegetation of the North Temperate Zone, in which a high proportion of the world's biologists have been trained, is not floristically rich, i.e. it does not contain a very large number of species. In general, as one travels southwards towards the equator, the number of species per unit area increases, and this is true whether one starts in the northern parts of North America, Europe or Asia.

It is difficult to illustrate this phenomenon with absolute figures, for the local conditions in any particular region play a large part in determining the number of species present. In Europe there are variously estimated to be 11 000 to 14 000 native vascular plant species, while in America north of Mexico (a much larger area) there are said to be about 16 000, but the latter figure includes naturalized aliens. By contrast the Amazon Basin, lying on the tropics but a much smaller area than either Europe or North America, is thought to possess about 20 000 species. Within North America, British Columbia has approximately 2 500 native species, while California, also on the Pacific coast but less than half the size and 2 400 km further south, has about twice that number. Much further south, overlying the equator, the countries of Ecuador and Colombia combined, with an area similar to that of British Columbia, possess about 3 000 species of orchids alone.

In Europe, Britain possesses approximately 1 600 native species of vascular plant (excluding agamospecies), a figure easily exceeded by the French Mediterranean Département of Var, which is about the size of an average English county. France, nearly three times the size of Britain, has about 4 400 species and Spain, slightly smaller than France, about 4 900. To take three islands, Iceland has about 500 species, Ireland (slightly smaller in area) has about 1 100 and Sicily (less than half the size of Ireland) about 2 300. The Iberian, Italian and Balkan Peninsulas each possess approximately 5 000 native species, about the same number as California, which is of the same order of size and at similar latitudes.

There are probably three main reasons for the greater number of species nearer the equator. Firstly, many species are not adapted to growth in cold conditions, especially where frost occurs; secondly, there are often habitats for cold-requiring species in generally warm areas (on mountains), but not *vice versa*; and thirdly, cool temperate regions have been subjected quite recently to extensive glaciations, which must have caused the disappearance from those areas of many species which were unable to return after glaciation.

Reliable figures are available for few tropical areas, but there is no doubt that they are in general richer than even the Warm Temperate Zone, although there are exceptions. One of these, West Tropical Africa, as defined in the *Flora of West Tropical Africa*, possesses only about 6 000 species of vascular plants, a figure only 20% higher than that for the Iberian Peninsula, which is one-eighth the area. The low figure for that part of the tropics is probably explained by the relatively uniform topography. Overall, however, probably about two-thirds of the world's flowering plant species are tropical, i.e. about 155 000 out of 240 000. Of the former, approximately 30 000 are African, 35 000 Asian and Australasian, and 90 000 American. Probably around 11 000 of the world's 12 000 species of pteridophytes and 16 000 out of 23 000 bryophytes are tropical.[329]

A further complication in this concept of increasing richness towards the equator is the relatively narrow localization of the centres of genetic diversity (CGD) discussed in Chapter 7 (Fig. 7.4). These are all in tropical or warm-temperate zones, but there are large areas within these zones which are not particularly important as gene centres.

A study of 1 000 m² sample sites in America[151] gave the following average number of species of lianas and trees: rich temperate woodland in Missouri, 23; dry tropical forest in Venezuela and Costa Rica, 63; moist tropical forest in Panama and Brazil, 109; wet tropical forest in Ecuador and Panama, 143; and pluvial tropical forest in Colombia, 258. The number of species in the four tropical areas of increasing rainfall are respectively 2.7, 4.7, 6.2 and 11.2 times the Missouri figure. Detailed studies such as this undermine the now frequently stated belief that the number of species in the tropics, particularly the American tropics, has been over-estimated in the past. Such beliefs are mainly based upon the fact that as more tropical Floras are being produced the number of accepted species in the area tends to decrease, due to the reduction of some to synonymy. However, this is only one side of the coin. A great many new species are being described from the tropics, even from relatively well-worked areas. For example over 7 000 new species were described from Africa south of the Sahara between 1953 and 1973, and many more than this from tropical Asia and America. Moreover, as more detailed work is carried out *in situ* in tropical regions, a substantial number of species that were reduced to synonymy in recent Floras are being resurrected; field and experimental work has confirmed their distinctness, whereas herbarium-based studies suggested otherwise. In well-worked temperate areas many species are recognized that would certainly not be upheld as good species in a tropical area in which few or no experimental studies had been performed, e.g. the two watercresses *Nasturtium officinale* and *N. microphyllum*, and the 91 European microspecies in the *Festuca ovina* aggregate. Groups such as these must surely exist in large numbers in the tropics; they await discovery and incorporation into the statistics. It is important to remember that the number of new species being described per year, even in the flowering plants, has not decreased in the past 50 years. This is true even in well-studied areas such as Europe. Since Volume 1 of *Flora Europaea* appeared in 1964, about 215 new species have been described from Europe in the families covered in that volume, and about two-thirds of these will be included in the second edition (J. R. Akeroyd pers. comm.). This represents an increase in accepted native species of about 5.5% in just over 20 years.

Importance of rich floras

Since the CGDs hold what remains of the ancestral genetic material of most of the world's taxa, a detailed knowledge of the flora of these regions is vital. For the improvement of man's crops (interpreted in the broadest sense) he must primarily seek greater genetic diversity, although, of course, the contributions to be made from the genetic manipulation of the already available spectrum of variation and from improved conditions of cultivation must not be under-estimated. Genetic variation beyond that already in cultivation exists only in areas in which the crop is native, and most of it only within its CGD. Not only do these genetic centres harbour greater extremes of variation of known crop species, but they are also likely to possess further species which can be developed into new crops, for it is believed that early man's selection of crop species incorporated a considerable element of chance.

The study, exploitation and conservation of CGDs are now considered high priorities by such bodies as the Food and Agriculture Organization of the United Nations Organization (FAO), and much work and many publications are directed at them.[139,406] Traditionally, agricultural systems within CGDs have utilized a much wider spectrum of the total variation of crop species than have those systems outside, although in the latter areas the crops are mostly highly selected to produce a very economical, high yield. Within CGDs particular cultivars may be very localized, being grown in a small area and differing from cultivars grown in neighbouring areas. These primitive culti-vars (*land-races*) are not highly productive compared with the advanced cultivars (*pan-cultivars*) grown in the more developed countries of Europe, North America and Eastern Asia, but they are often very well adapted to the local conditions, which are very different from those of cool temperate regions. Nowadays, with the economic 'development' of the CGDs, the less adaptable but more productive modern pan-cultivars are rapidly ousting the land-races, many of which are becoming extinct. This involves a narrowing of the total available genetic variability of the crop plants (*gene erosion*), which results in a reduction in the possibilities of introducing new aspects of variation into our crop plants. Gene erosion is occurring not only by the abandonment of old cultivars, but by the far cleaner methods of agriculture now being practised. These involve purer seed samples, the increased use of fertilizers and weed-killers, and the clearance of marginal areas (*niche erosion*) which once supported a range of genetically related taxa frequently contributing (by hybridization) to the genetic bases of the crop.

Realization that reliance on a small number of pan-cultivars might not be desirable in the long term, even if it leads to higher productivity at present, has also led to the search for entirely new species of crop plants. *Under-exploited Tropical Plants with Promising Economic Value*, produced by the U.S. National Academy of Sciences (1975), is one of many publications with this objective. The economic advantages arising from the development of new, native crops in the under-developed countries are obvious. In Nigeria, for example, big advances are being made in the Indigenous Fruit Trees Project organized by the Forestry Commission.[312] There seems to be good cause for optimism that such programmes will reveal valuable new crops in many warmer parts of the world.

A further reason to concentrate study on CGDs is that we are most likely to formulate generally applicable theories on plant variation and evolution by studying areas which contain the greatest diversity of genotypes. It is a fact that, just as our crops are based on a narrow part of the spectrum of plant variation, so too is all our botanical knowledge. Moreover, inasmuch as the well-studied part of the spectrum is mostly North Temperate in origin, it is atypical of the whole.

There are still many aspects of plant evolution which can greatly benefit from investigation on a broader front,[422] and no grounds for considering that such investigation will not uncover important new principles of general application.[329,406]

Taxonomic priorities

From what has been stated above it is clear that there are two outstanding botanical priorities: the broadening of the basis of the study and utilization of plants, in particular towards the CGDs; and the conservation of genetic diversity in these areas, so that this broadened study can be undertaken and its results utilized. These two priorities are closely inter-related. Whereas conservation of rich floras is essential if they are to be studied in depth, they can be conserved judiciously only if we have enough information to make the right decisions. The two processes are mutually dependent and must proceed in parallel. They should be viewed against the background of three important considerations.

Firstly, this work is a most urgent priority, because the economic develop-ment of already domesticated rich areas as well as the devastation of hitherto untouched regions are continuing at an ever increasing rate. There are now sufficiently accurate predictions to confront us with the disturbing realization that almost the last of the natural tropical forest on the earth will have been cut down by the end of this century; by 1975 about 41.5% had been destroyed.[345] It perhaps needs emphasizing that 'the end of this century' is no longer a long way off and a problem for later generations, but only a decade away and a problem with which the *present* generation will be faced. In 1910, J. Huber reckoned that the Amazon forest contained about 20 000 species of flowering plant, only about half of which were scientifically described. It is clear that in that region, as well as in parts of tropical Asia, many species are being eradicated before they have been discovered by botanists. The number of these which would have been of value to man is, of course, unknown, but is unlikely to be negligible. The race is on to extract as much information as possible from tropical areas before the opportunity has been removed,[329,330] but the rate at which we are preparing an inventory compared with the rate of destruction is still 'pathetically low'.[345]

Secondly, the taxonomist has a key, perhaps the major, role to play in this race. For too long the priorities in conservation have been dictated by ecological principles alone, even though taxonomic principles are at least equally relevant. Areas in need of conservation on genetic grounds are by no means always the same as those in need of conservation for ecological reasons; indeed, areas of great genetic diversity are often disturbed, marginal habitats which can appear ecologically very unattractive. Programmes of conservation must aim at preserving genetic diversity (as defined by taxonomic data) as well as habitat diversity (indicated by ecological data). It is the taxonomist who is able to produce a synthesis of the results of the systematic examination of organisms, and it is his synthesis which should pinpoint the gaps in our knowledge and hence the priorities for conservation and future research. It should also be realized that although the conservation of genetic material in botanic gardens (as living plants) and gene-banks (as seed), far removed from the native areas of the plants, have their part to play, these measures are by no means substitutes for plant conservation *in situ*.[138] Removal of material for conservation elsewhere can only involve the preservation of a small sample of the genetic diversity and, unless it is preceded by a very thorough taxonomic and genetic investigation, there is no certainty that it will be a representative sample.

Thirdly, the present pattern of taxonomic research is not designed or destined to make best use of the short time available. By historical accident plant taxonomy arose in the species-poor North Temperate Zone, and its most vigorous persuance is still carried on there.[406] There is an inverse correlation between the richness of the floras and both the amount which we have discovered about them and the extent to which we are now studying them. This is true with respect to the tropics versus temperate regions, to the American tropics versus the Old World tropics[329] and to various temperate regions (e.g. southern versus northern Europe).

There are, of course, many problems associated with scientific work in the tropics—climate, diseases, supplies, communications, etc.—but greater attempts must be made to overcome them. Prance[329,330] listed some priorities for the taxonomic exploration of the tropics. These include a greater concentration on the collection of hitherto poorly preserved material, for example fruits of tall trees and climbers, and fleshy plants; the production of better data labels with the specimens; more emphasis on economically important plants and upon plants of secondary vegetation; more visits by specialists, rather than their relying on the specimens brought back by general collectors; more conscious attempts to concentrate on the collection and study of rare species; and the training of more local botanists and the setting up of more local herbaria, so that extended, detailed studies of small areas, involving biosystematic as well as orthodox taxonomic investigations, can be undertaken.

Floras versus monographs

The nature of most basic taxonomic manuals, viz. the Flora, does not provide the best vehicle for the presentation of systematic data on plant species. Despite this, far greater effort is nowadays expended in the production of Floras than of monographs. The Flora is superficially a more attractive proposition from utilitarian, economical and political points of view. Being confined to a particular region, it is well geared to the potential consumers who live there, so that it will enjoy reasonable sales and popularity and is therefore likely to attract the patronage of publishers, institutions and governments. Hence many Floras throughout the world are being financed at official levels. Examples are *Flora Malesiana* in Leiden, *Flora of Tropical East Africa* at Kew, *Flora Meso-Americana* at the Herbario Nacional in Mexico, British Museum and Missouri Botanical Garden, *Flora of Ecuador* in Göteborg and Stockholm, and *Flora Neotropica* in New York. Conversely, monographs are not sponsored at official levels and virtually all of those completed and in the process of being compiled are the undertakings of individuals or small groups, led on only by personal ambition and a belief that they are on the right lines.

Nevertheless, Floras have disadvantages compared with monographs in several vital aspects. In particular, monographs seek to consider the total variation of the taxa covered, not simply that encountered in a specific, often completely arbitrarily (e.g. politically) defined, region covered by a Flora. For this reason Floras in general are taxonomically less sound than monographs. They are also almost invariably less thorough, since they concentrate very heavily on providing a means of identification of the species present in the region. Monographs generally also provide a means of identification, but they are primarily a systematic synthesis of the available data on the taxon in question. Such topics as biosystematic information and infraspecific variation are therefore dealt with at length in monographs, yet often completely neglected in Floras. These types of data are certainly within the terms of reference of Critical Floras, but the latter do not yet exist in reasonable numbers and are in any case better regarded as regional monographs. Furthermore, Floras are mostly very conservative, since they are aimed at providing a standard point of reference rather than presenting the results of new taxonomic research. Monographs are far more innovative, and in themselves represent real taxonomic advances. A random sample of 20 modern Floras (all published since 1950) with respect to the systematic sequence of angiosperms which was adopted showed that two were arranged alphabetically, two after various authorities, two after Bentham and Hooker (1862–1883), one after Boissier (1867–1888), three after Hutchinson (1926, 1959), one after Engler (1936), one after Emberger (1960), seven after Engler (1964) and one used a novel sequence. However, all but one of these systems are effectively pre-war, since the later Hutchinson and Engler schemes are largely based on their previous versions and the novel scheme on that of Bentham and Hooker; Emberger's unique system has no proponents outside France.

Radically new thinking, and new arguments, are needed if the current fashion for producing Floras rather than monographs is to be reversed.

Hitherto the purely academic arguments favouring the latter have been outweighed by the utilitarian, financial and political arguments referred to above. The facts that there are far more widely applicable and useful data contained in monographs, that the preparation of Floras of neighbouring regions often involves much wastefully repetitive work on the same taxa, that the writing of Floras would be far more rapid and easy if good monographs were available, and that the need for wide-scale monographing of plant taxa is extremely urgent, need to be brought forcibly to the attention of all those in relevant authority.

The need for new monographs

Clearly, the ideal goal is a monograph of all plant groups for the whole world, down to the lowest infraspecific levels; equally clearly, such a goal is unrealistic. It is therefore a case of determining a compromise at which to aim, based on our taxonomic priorities. Our knowledge of vascular plants greatly exceeds that of lower plants, and it is probably true that for the latter we have insufficient data and too few taxonomists to contemplate a new monographic phase. For vascular plants the view held by the writer is that a programme of monograph preparation is possible and desirable, and should follow two paths: a complete inventory down to the species level; and a selective detailed treatment down to the lowest levels. These two schemes are discussed below.

The complete inventory of species would amount to a new *Species Plantarum*—an alpha-taxonomic treatment of the world's species at the revisional level. For too long botanists have said that such a scheme is impossible, because of the vast numbers of species involved, without examining the case. Some simple arithmetic will present the facts. Since 1753 about 1 million species names of vascular plants have been published, representing an estimated 250 000 species. It would surely not be unreasonable to suggest that 500 taxonomists working on a ten-year programme could produce a revision of all these taxa. This would imply a loading of 500 species and 2 000 binomials per taxonomist, or 50 species and 200 binomials per taxonomist per year, or one species per week and less than one binomial per working day per taxonomist throughout the period. A great many species and binomials have in fact already been sufficiently dealt with in Floras and monographs, so would require relatively little work.

The costs involved are also calculable. Assuming a salary of £20 000 per annum per taxonomist, the total salary bill would amount to £100 million over the ten-year period. At the international level this sum is trifling. Moreover the requisite taxonomists and their institutional facilities are already in existence. It would not require the provision of new taxonomists, laboratories and herbaria, but a change in emphasis towards a common goal of the work carried out at the moment in the world's leading taxonomic institutions. Such a scheme would require a monumental international human commitment which is, to say the least, unlikely. Nevertheless, its completion would represent a wonderful achievement, and it should be made clear that our failure to produce it is not a measure of its intrinsic difficulty but of our lack of ability to work together with a single purpose. There is no doubt that if the world's politicians were to be convinced that the completion of a new *Species Plantarum* were a high priority, it would be produced. To deny this is much more short-sighted than denying, in 1959, that man would step on the moon ten years later, because in our example the costs are vastly less and the technology and facilities already exist. In fact a comparable undertaking to this has recently been launched in the form of the Human Genome Project, a plan to map completely the DNA of *Homo sapiens*, with a budget of ten times that mentioned above.

The infraspecific levels of the taxonomic hierarchy are in a state of great confusion, and we are not in a position to produce an inventory for more than a very small fraction of the world's flora. Not only have we not examined sufficiently thoroughly infraspecific variation in a wide range of species, but we do not have the means to express what we have discovered in taxonomic terms (see Chapter 8). For this reason a selective treatment, using certain groups of plants as examples, would be necessary in the first instance, so that new methods of infraspecific classification, with new rules of nomenclature, could be tried out. The abandonment of the lower levels of the hierarchy in many modern publications, pointed out in Chapter 8, is a trend which must be reversed by the provision of an acceptable methodology, for the cataloguing of infraspecific variation is undoubtedly of great importance. Groups which could be used in trial schemes include relatively small, well-known floras (e.g. Britain, Iceland), biosystematically well researched taxa (e.g. *Geum*, *Crepis*), and important crop-plants (e.g. *Brassica*, *Triticum*).

It is not logical to claim that one of these priorities is of higher standing than the other, any more than it is possible to claim that either the brain or the heart is more important to the human body; both are vital and we must strive towards both of them in parallel.

Conclusions

The following are here considered among the most important features of plant taxonomy which ideally should be concentrated upon towards the end of this century.

1   The further broadening of the application of a very wide range of taxonomic characters to all groups of plants, and in particular to:

    a   non-vascular plants
    b   the tropics
    c   centres of genetic diversity.

2   The investigation of a series of selected groups of plants by both phenetic and cladistic methods, and the comparison of the results obtained with those from 'narrative' methods, to formulate generalized views on the relative merits of the three approaches.

3   The compilation of an inventory of plants (of revisional status) down to the species level, i.e. a reversal of the current fashion so that there is a return to monographic rather than floristic work.

4   The formulation of a new, acceptable system of infraspecific classification which should be worked out on selected, well studied taxa and lead to the production of a new style of Critical Flora (regional monograph).

5   The application of taxonomic data to formulate plans of action in the conservation of genetic diversity *in situ* (as well as in gene banks, etc.).

Only if plant taxonomy is pursued intelligently, bearing in mind its impact upon and the requirements of its consumers (i.e. those who wish to use the classifications produced), will it be regarded as the barometer of the state of our total knowledge of plants, and as the means of directing future research, which unique position it must surely hold.

Appendix

Outline classification of the living plant kingdom (excluding procaryotes, fungi and lichens) down to the rank of class (in some cases subclass), with some common vernacular names.

CLASSIFICATION	VERNACULAR SYNONYMS	VERNACULAR GROUPINGS
Kingdom		
Subkingdoms		
Divisions		
Subdivisions		
Classes		
Subclasses		
Plantae		
Thallobionta	Thallophytes	
Rhodophyta	Red algae	
Rhodophyceae		
Cryptophyta	Cryptomonads	
Cryptophyceae		
Dinophyta (Pyrrophyta)	Dinoflagellates	
Dinophyceae		
Desmophyceae		
Xanthophyta		
Chloromonadophyceae	Chloromonads	
Xanthophyceae	Yellow-green algae	
Eustigmatophyceae		
Chrysophyta		
Chrysophyceae	Golden-brown algae	
Haptophyceae	Coccolithophorids	Algae
Bacillariophyta	Diatoms	
Bacillariophyceae		
Phaeophyta	Brown algae	
Phaeophyceae		

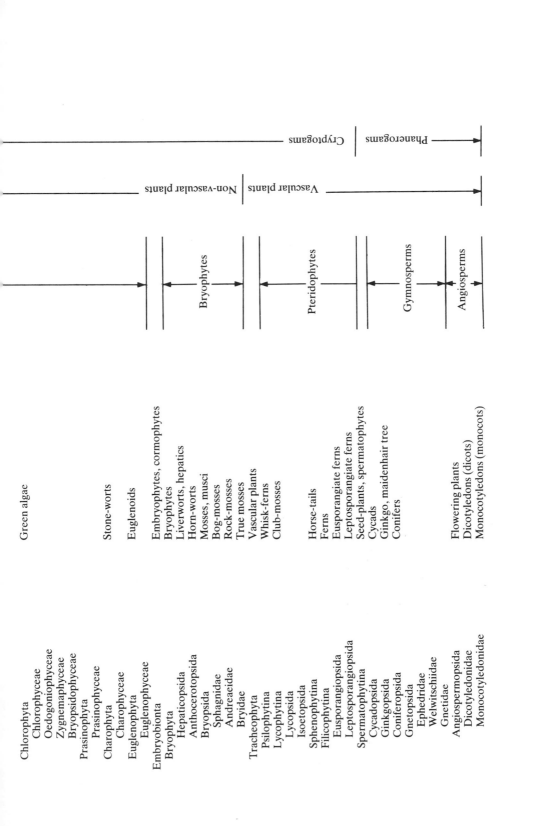

References

1 Abbott, L.A., Bisby, F.A. and Rogers, D.J. (1985). *Taxonomic Analysis in Biology. Computers, Models and Databases*. Columbia University Press, New York.
2 Abdallah, M.S. and De Wit, H.C.D. (1967–1978). *The Resedaceae. A taxonomical revision of the family*. Veenman & Zonen, Wageningen.
3 Abdulrahman, F.S. and Winstead, J.E. (1977). Chlorophyll levels and leaf ultrastructure as ecotypic characters in *Xanthium strumarium* L. *Am. J. Bot.*, **64**, 1177–1181.
4 Adams, R.P. and Turner, B.L. (1970). Chemosystematic and numerical studies of natural populations of *Juniperus ashei* Buch. *Taxon*, **19**, 728–751.
5 Alston, R.E., Rösler, H., Naifeh, K. and Mabry, T.J. (1965). Hybrid compounds in natural interspecific hybrids. *Proc. natn. Acad. Sci. U.S.A.*, **54**, 1458–1465.
6 Alston, R.E. and Turner, B.L. (1963a). *Biochemical Systematics*. Prentice-Hall, New Jersey.
7 Alston, R.E. and Turner, B.L. (1963b). Natural hybridization among four species of *Baptisia* (Leguminosae). *Am. J. Bot.*, **50**, 159–173.
8 Anderson, E. (1948). Hybridization of the habitat. *Evolution, Lancaster, Pa.*, **2**, 1–9.
9 Anderson, E. (1949). *Introgressive Hybridization*. Wiley, New York.
10 Anderson, E. and Hubricht, L. (1938). Hybridization in *Tradescantia*, 3. The evidence for introgressive hybridization. *Am. J. Bot.*, **25**, 396–402.
11 Anonymous (1960). *Identification of Hardwoods. A Lens Key. Forest Products Research Bulletin*, **25**. Her Majesty's Stationery Office, London.
12 Arber, A. (1938). *Herbals: Their Origin and Evolution*. 2nd edition. Cambridge University Press.
13 Babcock, E.B. (1947). The genus *Crepis*, Parts I and II. *Univ. Calif. Publs Bot.*, **21** and **22**.
14 Baker, H.G. and Baker, I. (1973). Some anthecological aspects of the evolution of nectar-producing flowers, particularly amino acid production in nectar. In *Taxonomy and Ecology*, Heywood, V.H. (ed.), 243–264. Academic Press, London and New York.
15 Barber, H.N. (1970). Hybridization and the evolution of plants. *Taxon*, **19**, 154–160.
16 Barker, C.M. and Stace, C.A. (1986). Hybridization in the genera *Vulpia* and *Festuca* (Poaceae): meiotic behaviour of artificial hybrids. *Nordic J. Bot.*, **6**, 1–10.
17 Barthlott, W. (1984). Microstructural features of seed surfaces. In *Current Concepts in Plant Taxonomy*, Heywood, V.H. and Moore, D.M. (eds), 95–105. Academic Press, London.
18 Bate-Smith, E.C. (1962). The phenolic constituents of plants and their taxonomic significance. *J. Linn. Soc. (Bot.)*, **58**, 95–173.

19 Baum, B.R. (1977). *Oats: Wild and Cultivated. A Monograph of the genus Avena L. (Poaceae)*. Minister of Supply and Services, Ottawa.
20 Behnke, H.-D. (1975). The bases of angiosperm phylogeny: Ultrastructure. *Ann. Mo. bot. Gdn*, **62**, 647–663.
21 Behnke, H.-D. (1977). Dilatierte ER-Zisternen, ein mikromorphologisches Merkmal des Capparales? *Ber. dt. bot. Ges.*, **90**, 241–252.
22 Behnke, H.-D. (ed.) (1981). Ultrastructure and systematics of seed plants. *Nordic J. Bot.*, **1**, 341–460.
23 Behnke, H.-D. and Barthlott, W. (1983). New evidence from the ultrastructural and micromorphological fields in angiosperm classification. *Nordic J. Bot.*, **3**, 43–66.
24 Belford, H.S. and Thompson, W.F. (1981). Single copy DNA homologies in *Atriplex*. I. Cross reactivity estimates and the role of deletions in genome evolution. *Heredity*, **46**, 91–108. II. Hybrid thermal stabilities and molecular phylogeny. *Heredity*, **46**, 109–122.
25 Bendz, G. and Santesson, J. (eds) (1974). *Chemistry in Botanical Classification. Nobel Symposium*, **25**. Academic Press, London and New York.
26 Bennell, A.P. and Henderson, D.M. (1985). Rusts and other fungal parasites as aids to pteridophyte taxonomy. *Proc. R. Soc. Edinb.*, B, **86**, 115–124.
27 Bennett, M.D. (1984). The genome, the natural karyotype, and biosystematics. In *Plant Biosystematics*, Grant, W.F. (ed.), 41–66. Academic Press, Toronto.
28 Bennett, M.D. and Smith, J.B. (1976). Nuclear DNA amounts in angiosperms. *Phil. Trans. R. Soc. Lond.*, Ser. B, **274**, 227–274.
29 Bennett, M.D., Smith, J.B. and Heslop-Harrison, J.S. (1982). Nuclear DNA amounts in angiosperms. *Proc. R. Soc. Lond.*, Ser. B, **216**, 179–199.
30 Benoit, P.M. (1961). *Catapodium marinum × rigidum. Proc. bot. Soc. Br. Isl.*, **4**, 276.
31 Benson, L.D. (1962). *Plant Taxonomy: Methods and Principles*. Ronald Press, New York.
32 Berlin, B. (1973). Folk systematics in relation to biological classification and nomenclature. *A. Rev. Ecol. Syst.*, **4**, 259–271.
33 Berlin, B., Breedlove, D.E. and Raven, P.H. (1973). General principles of classification and nomenclature in folk biology. *Am. Anthrop.*, **75**, 214–242.
34 Bessey, C.E. (1915). The phylogenetic taxonomy of the angiosperms. *Ann. Mo. bot. Gdn*, **2**, 109–164.
35 Birch, A.J. (1963). Biosynthetic pathways. In *Chemical Plant Taxonomy*, Swain, T. (ed.), 141–166. Academic Press, London and New York.
36 Birks, H.J.B. and Deacon, J. (1973). A numerical analysis of the past and present flora of the British Isles. *New Phytol.*, **72**, 877–902.
37 Bisby, F.A., Vaughan, J.G. and Wright, C.A. (eds) (1980). *Chemosystematics: Principles and Practice*. Academic Press, London.
38 Blackmore, S. (1984). Pollen features and plant systematics. In *Current Concepts in Plant Taxonomy*, Heywood, V.H. and Moore, D.M. (eds), 135–154. Academic Press, London.
39 Blakeslee, A.F., Avery, A.G., Satina, S. and Rietsama, J. (1959). *The Genus Datura*. Ronald Press, New York.
40 Blunt, W. (1971). *The Compleat Naturalist. A Life of Carl Linnaeus*. Collins, London.
41 Böcher, T.W. (1972). Evolutionary problems in the Arctic flora. In *Taxonomy, Phytogeography and Evolution*, Valentine, D.H. (ed.), 101–113. Academic Press, London and New York.
42 Bolkhovskikh, Z., Grif, V., Matvejeva, T. and Zakharyeva, O. (1969). *Chromosome Numbers of Flowering Plants*. Academy of Sciences U.S.S.R., Leningrad.

43   Boulter, D. (1974). The use of amino acid sequence data in the classification of higher plants. In *Chemistry in Botanical Classification. Nobel Symposium*, **25**, Bendz, G. and Santesson, J. (eds), 211–216. Academic Press, London and New York.

44   Boulter, D., Peacock, D., Guise, A., Gleaves, J.T. and Estabrook, G. (1979). Relationships between the partial amino-acid sequences of plastocyanin from members of ten families of flowering plants. *Phytochemistry*, **18**, 603–608.

45   Boulter, D., Ramshaw, J.A.M., Thompson, E.W., Richardson, M. and Brown, R.H. (1972). A phylogeny of higher plants based on the amino acid sequence of cytochrome *c* and its biological implications. *Proc. R. Soc. Lond.*, Ser. B, **181**, 441–455.

46   Bradshaw, A.D. (1962). The taxonomic problems of local geographical variation in plant species. In *Taxonomy and Geography*, Nichols, D. (ed.), 7–16. Systematics Association, London.

47   Bramwell, D. (1972). Endemism in the flora of the Canary Islands. In *Taxonomy, Phytogeography and Evolution*, Valentine, D.H. (ed.), 141–159. Academic Press, London and New York.

48   Brazier, J.D. and Franklin, G.L. (1961). *Identification of Hardwoods. A Microscope Key. Forest Products Research Bulletin*, **46**. Her Majesty's Stationery Office, London.

49   Bremer, K. and Wanntorp, H.-E. (1978). Phylogenetic systematics in botany. *Taxon*, **27**, 317–329.

50   Bremer, K. and Wanntorp, H.-E. (1981). The cladistic approach to plant classification. In *Advances in Cladistics*, **1**, Funk, V.A. and Brooks, D.R. (eds), 87–94. New York Botanical Garden, New York.

51   Brewbaker, J.L. (1967). The distribution and phylogenetic significiance of binucleate and trinucleate pollen grains in the angiosperms. *Am. J. Bot.*, **54**, 1069–1083.

52   Briggs, D. and Block, M. (1981). An investigation into the use of the '-deme' terminology. *New Phytol.*, **89**, 729–735.

53   Briggs, D. and Walters, S.M. (1984). *Plant Variation and Evolution*. 2nd edition. Cambridge University Press.

54   Brooks, D.R., Caira, J.N., Platt, T.R. and Pritchard, M.R. (1984). *Principles and Methods of Phylogenetic Systematics: a Cladistics Workbook*. University of Kansas Museum of Natural History, Lawrence, Kansas.

55   Buch, H. (1922–1928). Die Scapanien Nordeuropas und Sibiriens, 1 & 2. *Commentat. biol.*, **1(4)** and **3(1)**.

56   Burnett, W.C., Jones, S.B. and Mabry, T.J. (1977). Evolutionary implications of sesquiterpene lactones in *Vernonia* (Compositae) and mammalian herbivores. *Taxon*, **26**, 203–207.

57   Cadbury, D.A., Hawkes, J.G. and Readett, R.C. (1971). *A Computer-Mapped Flora. A Study of the County of Warwickshire*. Academic Press, London and New York.

58   Cain, A.J. (ed.) (1959). *Function and Taxonomic Importance*. Systematics Association, London.

59   Camin, J.H. and Sokal, R.R. (1965). A method for deducing branching sequences in phylogeny. *Evolution, Lancaster, Pa.*, **19**, 311–326.

60   Camp, W.H. and Gilly, C.L. (1943). The structure and origin of species. *Brittonia*, **4**, 323–385.

61   Chesters, K.I.M., Gnauck, F.R. and Hughes, N.F. (1967). Angiospermae. In *The Fossil Record*, Harland, W.B. *et al.* (eds), 269–288. The Geological Society, London.

62   Christie, A., Pocock, K., Lewis, D.H. and Duckett, J.G. (1985). A comparison between the carbohydrates of axenically cultured hepatics and those collected from the field. *J. Bryol.*, **13**, 417–421.

63 Clarke, G.C.S. (1979). Spore morphology and bryophyte systematics. In *Bryophyte Systematics*, Clarke, G.C.S. and Duckett, J.G. (eds), 231–250. Academic Press, London.

64 Clarke, S.H. (1938). A multiple-entry perforated-card key with special reference to the identification of hardwoods. *New Phytol.*, **37**, 369–374.

65 Clausen, J., Keck, D.D. and Hiesey, W.M. (1940). Experimental studies on the nature of species, I. The effect of varied environments on western North American plants. *Publs Carnegie Instn*, **520**.

66 Clausen, J., Keck, D.D. and Hiesey, W.M. (1945). Experimental studies on the nature of species, II. Plant evolution through amphiploidy and autoploidy, with examples from the Madiinae. *Publs Carnegie Instn*, **564**.

67 Clausen, J., Keck, D.D. and Hiesey, W.M. (1948). Experimental studies on the nature of species, III. Environmental responses of climatic races of *Achillea*. *Publs Carnegie Instn*, **581**.

68 Cleland, R.E. (1972). Oenothera: *Cytogenetics and Evolution*. Academic Press, London and New York.

69 Cockburn, W. (1985). Variation in photosynthetic acid metabolism in vascular plants: CAM and related phenomena. *New Phytol.*, **101**, 3–24.

70 Cook, C.D.K. (1968). Phenotypic plasticity with particular reference to three amphibious plant species. In *Modern Methods in Plant Taxonomy*, Heywood, V.H. (ed.), 97–111. Academic Press, London and New York.

71 Cook, C.D.K. (1970). Hybridization in the evolution of *Batrachium*. *Taxon*, **19**, 161–166.

72 Core, E.L. (1955). *Plant Taxonomy*. Prentice-Hall, Englewood Cliffs.

73 Corner, E.J.H. (1976). *The Seeds of Dicotyledons*, **1** and **2**. Cambridge University Press.

74 Crisci, J.V. and Stuessy, T.F. (1980). Determining primitive character states for phylogenetic reconstruction. *Syst. Bot.*, **5**, 112–135.

75 Cotton, R. and Stace, C.A. (1977). Morphological and anatomical variation of *Vulpia* (Gramineae). *Bot. Notiser*, **130**, 173–187.

76 Croizat, L., Nelson, G. and Rosen, D.E. (1974). Centers of origin and related concepts. *Syst. Zool.*, **23**, 265–287.

77 Cronquist, A. (1981). *An Integrated System of Classification of Flowering Plants*. Columbia University Press, New York.

78 Cronquist, A. (1987). A botanical critique of cladism. *Bot. Rev.*, **53**, 1–52.

79 Crundwell, A.C. (1985). The introduced bryophytes of the British Isles. *Bull. Br. bryol. Soc.*, **45**, 8–9.

80 Crundwell, A.C. and Nyholm, E. (1964). The European species of the *Bryum erythrocarpum* complex. *Trans. Br. bryol. Soc.*, **4**, 597–637.

81 Culberson, C.F. (1969). *Chemical and Botanical Guide to Lichen Products*. University of Carolina Press, Chapel Hill.

82 Culberson, W.L. (1970). Chemosystematics and ecology of lichen-forming fungi. *A. Rev. Ecol. Syst.*, **1**, 153–170.

83 Culberson, W.L. and Culberson, C.F. (1970). A phylogenetic view of chemical evolution in the lichens. *Bryologist*, **73**, 1–31.

84 Cullen, J. (1978). A preliminary survey of ptyxis (vernation) in the angiosperms. *Notes R. bot. Gdn Edinb.*, **37**, 161–214.

85 Cutler, D.F. (1972). Vicarious species of Restionaceae in Africa, Australia and South America. In *Taxonomy, Phytogeography and Evolution*, Valentine, D.H. (ed.), 73–83. Academic Press, London and New York.

86 Cutler, D.F. (1984). Systematic anatomy and embryology—recent develop-

ments. In *Current Concepts in Plant Taxonomy*, Heywood, V.H. and Moore, D.M. (eds), 107–133. Academic Press, London.

87   Czaja, A.T. (1978). Structure of starch grains and the classification of vascular plant families. *Taxon*, **27**, 463–470.

88   Daday, H. (1954). Gene frequencies in wild populations of *Trifolium repens*, 1. Distribution by latitude. *Heredity, Lond.*, **8**, 61–78.

89   Dahlgren, R. (1975a). A system of classification of the angiosperms to be used to demonstrate the distribution of characters. *Bot. Notiser*, **128**, 119–147.

90   Dahlgren, R. (1975b). The distribution of characters within an angiosperm system, I. Some embryological characters. *Bot. Notiser*, **128**, 181–197.

91   Dahlgren, R. (1977). A commentary on a diagrammatic presentation of the angiosperms in relation to the distribution of character states. *Pl. Syst. Evol.*, Suppl. **1**, 253–283.

92   Dahlgren, R. (1983). General aspects of angiosperm evolution and macrosystematics. *Nordic J. Bot.*, **3**, 119–149.

93   Dahlgren, R., Clifford, H.T. and Yeo, P.F. (1985). *The Families of the Monocotyledons: Structure, Evolution and Taxonomy*. Springer-Verlag, Berlin.

94   Danser, B.H. (1929). Über die Begriffe Komparium, Kommiscuum und Konvivium und über die Entstehungsweise der Konvivien. *Genetica*, **11**, 399–450.

95   Darlington, C. D. (1937). What is a hybrid? *J. Hered.*, **28**, 308.

96   Daussant, J. and Skakoun, A. (1983). Immunochemistry of seed proteins. In *Seed Proteins*, Daussant, J., Mossé, J. and Vaughan, J. (eds), 101–133. Academic Press, London.

97   Davis, G.L. (1966). *Systematic Embryology of the Angiosperms*. Wiley, New York.

98   Davis, P.H. and Heywood, V.H. (1963). *Principles of Angiosperm Taxonomy*. Oliver & Boyd, Edinburgh and London.

99   Diamond, J.M. (1983). Taxonomy by nucleotides. *Nature, Lond.*, **305**, 17–18.

100 Dickison, W.C. (1975). The bases of angiosperm phylogeny: Vegetative anatomy. *Ann. Mo. bot. Gdn*, **62**, 590–620.

101 Dilcher, D.L. (1974). Approaches to the identification of angiosperm leaf remains. *Bot. Rev.*, **40**, 1–157.

102 Donoghue, M.J. and Maddison, W.P. (1986). Polarity assessment in phylogenetic systematics: a response to Meacham. *Taxon*, **35**, 534–538.

103 Duncan, T. (1980). Cladistics for the practising taxonomist—an eclectic view. *Syst. Bot.*, **5**, 136–148.

104 Duncan, T. and Meacham, C.A. (1986). Multiple-entry keys for the identification of angiosperm families using a microcomputer. *Taxon*, **35**, 492–494.

105 Duncan, T., Phillips, R.B. and Wagner, W.H. jr (1980). A comparison of branching diagrams derived by various phenetic and cladistic methods. *Syst. Bot.*, **5**, 264–293.

106 Du Rietz, G.E. (1930). The fundamental units of biological taxonomy. *Svensk bot. Tidskr.*, **24**, 333–428.

107 Ebinger, J.E. (1962). *Luzula × borreri* in England. *Watsonia*, **5**, 251–254.

108 Edmonds, J.M. (1978). Numerical taxonomic studies on *Solanum* L. section *Solanum (Maurella)*. *Bot. J. Linn. Soc.*, **76**, 27–51.

109 Edwards, P. (1976). A classification of plants into higher taxa based on cytological and biochemical criteria. *Taxon*, **25**, 529–542.

110 Ehrendorfer, F. (1959). Differentiation-hybridization cycles and polyploidy in *Achillea*. *Cold Spring Harbor Symp. Quant. Biol.*, **24**, 141–152.

111 Ehrendorfer, F. (1968). Geographical and ecological aspects of infraspecific differentiation. In *Modern Methods in Plant Taxonomy*, Heywood, V.H. (ed.), 261–296. Academic Press, London and New York.

112 Ehrendorfer, F. (1973). Adaptive significance of major taxonomic characters and morphological trends in angiosperms. In *Taxonomy and Ecology*, Heywood, V.H. (ed.), 317–327. Academic Press, London and New York.

113 Ehrendorfer, F. (1983). Quantitative and qualitative differentiation of nuclear DNA in relation to plant systematics and evolution. In *Proteins and Nucleic Acids in Plant Systematics*, Jensen, U. and Fairbrothers, D.E. (eds), 3–35. Springer-Verlag, Berlin.

114 Ehrendorfer, F. (1986). Chromosomal differentiation and evolution in angiosperm groups. In *Modern Aspects of Species*, Iwatsuki, K., Raven, P.H. and Bock, W.J. (eds), 59–86. University of Tokyo Press, Tokyo.

115 Eldredge, N. and Cracraft, J. (1980). *Phylogenetic Patterns and the Evolutionary Process*. Columbia University Press, New York.

116 Ellis, R.P. (1976). A procedure for standardizing comparative leaf anatomy in the Poaceae; I. The leaf-blade as viewed in transverse section. *Bothalia*, **12**, 65–109.

117 Ellis, R.P. (1979). A procedure for standardizing comparative leaf anatomy in the Poaceae, II. The epidermis as seen in surface view. *Bothalia*, **12**, 641–671.

118 Erdtman, G., Berglund, B. and Praglowski, J. (1961). *An Introduction to a Scandinavian Pollen Flora*. Almqvist & Wiksell, Stockholm.

119 Estabrook, G.F. (1978). Some concepts for the estimation of evolutionary relationships in systematic botany. *Syst. Bot.*, **3**, 146–158.

120 Estabrook, G.F. (1980). The compatibility of occurrence patterns of chemicals in plants. In *Chemosystematics: Principles and Practice*, Bisby, F.A., Vaughan, J.G. and Wright, C.A. (eds), 370–397. Academic Press, London.

121 Evans, D.A. (1983). Protoplast fusion. In *Handbook of Plant Cell Culture*, Evans, D.A., Sharp, W.R., Ammirato, P.V. and Yamada, Y. (eds), 291–321. Macmillan, New York.

122 Exell, A.W. (1960). Systematics Association Committee for descriptive terminology, I. Preliminary list of works relevant to descriptive biological terminology. *Taxon*, **9**, 245–257.

123 Exell, A.W. (1962). Systematics Association Committee for descriptive terminology, II. Terminology of simple symmetrical plane shapes (Chart 1). *Taxon*, **11**, 145–156.

124 Eyde, R.H. (1975). The bases of angiosperm phylogeny: Floral anatomy. *Ann. Mo. bot. Gdn*, **62**, 521–537.

125 Faegri, K. and Iversen, J. (1975). *Textbook of Pollen Analysis*. 3rd edition. Blackwell, Oxford and London.

126 Fairbrothers, D.E. (1969). Comparisons of proteins obtained from diverse plant organs for chemosystematic research. *Revue roum. Biochim.*, **6**, 95–103.

127 Fairbrothers, D.E. (1977). Perspectives in plant serotaxonomy. *Ann. Mo. bot. Gdn*, **64**, 147–160.

128 Fairbrothers, D.E. (1983). Evidence from nucleic acid and protein chemistry, in particular serology, in angiosperm classification. *Nordic J. Bot.*, **3**, 35–41.

129 Fairbrothers, D.E., Mabry, T.J., Scogin, R.L. and Turner, B.L. (1975). The bases of angiosperm phylogeny: Chemotaxonomy. *Ann. Mo. bot. Gdn*, **62**, 765–800.

130 Fairbrothers, D.E. and Petersen, F.P. (1983). Serological investigation of the Annoniflorae. In *Proteins and Nucleic Acids in Plant Systematics*, Jensen, U. and Fairbrothers, D.E. (eds), 301–310. Springer-Verlag, Berlin.

131 Favarger, C. (1972). Endemism in the montane floras of Europe. In *Taxonomy, Phytogeography and Evolution*, Valentine, D.H. (ed.), 191–204. Academic Press, London and New York.

132   Favarger, C. and Contandriopoulis, J. (1961). Essai sur l'endemisme. *Ber. schweiz. bot. Ges.*, **71**, 384–406.

133   Fernandes, A. (1931). Estudios nos cromosomas das Liliáceas e Amarilidáceas. *Bol. Soc. broteriana*, Sér. II, **7**, 1–122.

134   Fischer, M. (1975). The *Veronica hederifolia* group: taxonomy, ecology, and phylogeny. In *European Floristic and Taxonomic Studies*, Walters, S.M. and King, C.J. (eds), 48–60. E. W. Classey, Faringdon.

135   Flake, R.H., von Rudloff, E. and Turner, B.L. (1969). Quantitative study of clinal variation in *Juniperus virginiana* using terpenoid data. *Proc. natn. Acad. Sci. U.S.A.*, **64**, 487–494.

136   Florin, R. (1931). Untersuchungen zur Stammesgeschichte der Coniferales und Cordaitales. Erster Teil: Morphologie und Epidermisstruktur der Assimilationsorgane bei den rezenten Koniferen. *K. svenska VetenskAkad. Handl.*, Ser. 3, **10(1)**.

137   Florin, R. (1933). Studien über die Cycadales des Mesozoikums nebst Eröterungen über die Spaltöffnungsapparate der Bennettitales. *K. svenska Vetensk-Akad. Handl.*, Ser. 3, **12(5)**.

138   Frankel, O.H. (1970). Genetic conservation in perspective. In *Genetic Resources in Plants—Their Exploration and Conservation. I.B.P. Handbook*, **11**, Frankel, O.H. and Bennett, E. (eds), 469–489. Blackwell, Oxford and Edinburgh.

139   Frankel, O.H. and Bennett, E. (eds) (1970). *Genetic Resources in Plants—Their Exploration and Conservation. I.B.P. Handbook*, **11**. Blackwell, Oxford and Edinburgh.

140   Fritsch, R. (1982). *Index to Plant Chromosome Numbers—Bryophyta. Regnum veg.*, **108**. Bohn, Scheltema & Holkema, Utrecht and Antwerp, and Dr. W. Junk, The Hague and Boston.

141   Frodin, D.G. (1984). *Guide to Standard Floras of the World*. Cambridge University Press.

142   Fryxell, P.A. (1971). Phenetic analysis and the phylogeny of the diploid species of *Gossypium* L. (Malvaceae). *Evolution, Lancaster, Pa.*, **25**, 554–562.

143   Fukuda, I. (1984). Chromosome banding and biosystematics. In *Plant Biosystematics*, Grant, W.F. (ed.), 97–116. Academic Press, Toronto.

144   Funk, V.A. (1981). Special concerns in estimating plant phylogenies. In *Advances in Cladistics*, **1**, Funk, V.A. and Brooks, D.R. (eds), 73–86. New York Botanical Garden, New York.

145   Funk, V.A. (1985). Phylogenetic patterns and hybridization. *Ann. Mo. bot. Gdn*, **72**, 681–715.

146   Funk, V.A. and Stuessy, T.F. (1978). Cladistics for the practising plant taxonomist. *Syst. Bot.*, **3**, 159–178.

147   Gadgil, M. and Solbrig, O.T. (1972). The concept of *r*- and *K*-selection: evidence from wild flowers and some theoretical considerations. *Am. Nat.*, **106**, 14–31.

148   Gajewski, W. (1957). A cytogenetic study of the genus *Geum* L. *Monogr. bot.*, **4**, 1–416.

149   Gastony, G.J. (1977). Chromosomes of the independently reproducing Appalachian gametophyte: A new source of taxonomic evidence? *Syst. Bot.*, **2**, 43–48.

150   Geesink, R., Leeuwenberg, A.J.M., Ridsdale, C.E. and Veldkamp, J.F. (1981). *Thonner's Analytical Key to the Families of Flowering Plants*. Leiden Botanical Series, **5**. Leiden University Press, The Hague.

151   Gentry, A.H. (1982). Patterns of neotropical plant species diversity. *Evol. Biol.*, **15**, 1–84.

152   Gershenzon, J. and Mabry, T.J. (1983). Secondary metabolites and the higher classification of angiosperms. *Nordic J. Bot.*, **3**, 5–34.

153 Gibbs, R.D. (1974). *Chemotaxonomy of Flowering Plants*. McGill-Queen's University Press, Montreal.
154 Gill, J.J.B. and Walker, S. (1971). Studies on *Cytisus scoparius* (L.) Link with particular eference to the prostrate forms. *Watsonia*, **8**, 345–356.
155 Gillett, G.W. and Lim, E.K.S. (1970). An experimental study of the genus *Bidens* (Asteraceae) in the Hawaiian Islands. *Univ. Calif. Publs Bot.*, **56**, 1–63.
156 Gilmour, J.S.L. and Gregor, J.W. (1939). Demes: A suggested new terminology. *Nature, Lond.*, **144**, 333.
157 Gilmour, J.S.L. and Heslop-Harrison, J. (1954). The deme terminology and the units of microevolutionary change. *Genetica*, **27**, 147–161.
158 Godward, M.B.E. (ed.) (1966). *The Chromosomes of the Algae*. Edward Arnold, London.
159 Goodman, M.M. (1967). The identification of hybrid plants in segregating populations. *Evolution, Lancaster, Pa.*, **21**, 334–340.
160 Goodspeed, T.H. (1955). *The genus* Nicotiana. Chronica Botanica Co., Waltham.
161 Gottlieb, L.D. (1977). Electrophoretic evidence and plant systematics. *Ann. Mo. bot. Gdn*, **64**, 161–180.
162 Gottlieb, L.D. (1983). Isozyme number and phylogeny. In *Proteins and Nucleic Acids in Plant Systematics*, Jensen, U. and Fairbrothers, D.E. (eds), 209–221. Springer-Verlag, Berlin.
163 Grant, V. (1958). The regulation of recombination in plants. *Cold Spring Harbor Symp. Quant. Biol.*, **23**, 337–363.
164 Grant, V. (1971). *Plant Speciation*. Columbia University Press, New York.
165 Gray, J.C. (1980). Fraction I protein and plant phylogeny. In *Chemosystematics: Principles and Practice*, Bisby, F.A., Vaughan, J.G. and Wright, C.A. (eds), 167–193. Academic Press, London.
166 Gregor, J.W. (1939). Experimental taxonomy, IV. Population differentiation in North American and European Sea Plantains allied to *Plantago maritima* L. *New Phytol.*, **38**, 293–322.
167 Gregory, W.C. (1941). Phylogenetic and cytological studies in the Ranunculaceae. *Trans. Am. phil. Soc.*, **31**, 443–520.
168 Greilhuber, J. and Speta, F. (1978). Quantitative analysis of C-banded karyotypes, and systematics in the cultivated species of *Scilla siberica* group (Liliaceae). *Pl. Syst. Evol.*, **129**, 63–109.
169 Guinochet, M. (1973). Phytosociologie et systématique. In *Taxonomy and Ecology*, Heywood, V.H. (ed.), 121–140. Academic Press, London and New York.
170 Guinochet, M. and Vilmorin, R. de (1973). *Flore de France*, **1**. Centre National de la Recherche Scientifique, Paris.
171 Gustafsson, A. (1946). Apomixis in the higher plants, I. The mechanism of apomixis. *Acta Univ. lund.*, N.F. Avd. 2, **42(3)**, 1–66.
172 Gustafsson, A. (1947a). Apomixis in higher plants, II. The causal aspect of apomixis. *Acta Univ. lund.*, N.F. Avd. 2, **43(2)**, 71–178.
173 Gustafsson, A. (1947b). Apomixis in higher plants, III. Biotype and species formation. *Acta Univ. lund.*, N.F. Avd. 2, **44(2)**, 183–370.
174 Hall, M.T. (1952). Variation and hybridization in *Juniperus*. *Ann. Mo. bot. Gdn*, **39**, 1–64.
175 Halle, F., Oldeman, R.A.A. and Tomlinson, P.B. (1978). *Tropical Trees and Forests: an Architectural Analysis*. Springer-Verlag, Berlin.
176 Hamann, U. (1977). Über Konvergenzen bei embryologischen Merkmalen der Angiospermen. *Ber. dt. bot. Ges.*, **90**, 369–384.
177 Hara, H. (1972). Corresponding taxa in North America, Japan and the

Himalayas. In *Taxonomy, Phytogeography and Evolution*, Valentine, D.H. (ed.), 61–72. Academic Press, London and New York.

178   Harborne, J.B. (ed.) (1964). *Biochemistry of Phenolic Compounds*. Academic Press, London and New York.

179   Harborne, J.B. (1967). *Comparative Biochemistry of the Flavonoids*. Academic Press, London and New York.

180   Harborne, J.B. (ed.) (1970). *Phytochemical Phylogeny*. Academic Press, London and New York.

181   Harborne, J.B. (ed.) (1978). *Biochemical Aspects of Plant and Animal Coevolution*. Academic Press, London.

182   Harborne, J.B. (1984a). Chemical data in practical taxonomy. In *Current Concepts in Plant Taxonomy*, Heywood, V.H. and Moore, D.M. (eds), 237–261. Academic Press, London.

183   Harborne, J.B. (1984b). *Phytochemical Methods: a Guide to Modern Techniques of Plant Analysis*. 2nd edition. Chapman & Hall, London.

184   Harlan, J.R. (1971). Agricultural origins: centres and non-centres. *Science, N.Y.*, **174**, 468–474.

185   Haskell, G. (1966). The history, taxonomy and breeding system of apomictic British *Rubi*. In *Reproductive Biology and Taxonomy of Vascular Plants*, Hawkes, J.G. (ed.), 141–151. Pergamon Press, Oxford and London.

186   Hatheway, W.H. (1962). Weighted hybrid index. *Evolution, Lancaster, Pa.*, **16**, 1–10.

187   Hawkes, J.G. (ed.) (1968). *Chemotaxonomy and Serotaxonomy*. Academic Press, London and New York.

188   Hawkes, J.G. and Lester, R.N. (1968). Immunological studies on the tuber-bearing Solanums, III. Variability within *S. bulbocastanum. Ann. Bot.*, **32**, 165–186.

189   Hedge, I.C. and Lamond, J.M. (1969). A guide to the Turkish genera of Umbelliferae. *Notes R. bot. Gdn Edinb.*, **25**, 171–177.

190   Hedge, I.C. and Lamond, J.M. (1972). Umbelliferae. Multi-access key to the Turkish genera. In *Flora of Turkey*, **4**, Davis, P.H. (ed.), 280–288. Edinburgh University Press, Edinburgh.

191   Hegnauer, R. (1962–1973). *Chemotaxonomie der Pflanzen*, **1–6**, Birkhäuser, Basel.

192   Henderson, D.M. (ed.) (1983). *International Directory of Botanical Gardens IV*. Koeltz, Koenigstein.

193   Hennig, W. (1966). *Phylogenetic Systematics*. Translated by D. D. Davies and R. Zangerl. Univ. Illinois Press, Urbana.

194   Hennig, W. (1975). Cladistic analysis or cladistic classification: a reply to Ernst Mayr. *Syst. Zool.*, **24**, 244–256.

195   Henrey, B. (1975). *British Botanical and Horticultural Literature before 1800*, **1–3**. Oxford University Press.

196   Herter, G. (1949). *Index Lycopodiorum*. Privately printed. Montevideo.

197   Heslop-Harrison, J. (1953a). *New Concepts in Flowering Plant Taxonomy*. Heinemann, London.

198   Heslop-Harrison, J. (1953b). The North American and Lusitanian elements in the flora of the British Isles. In *The Changing Flora of Britain*, Lousley, J.E. (ed.), 105–123. Botanical Society of the British Isles, Oxford.

199   Heywood, V.H. (1959). The taxonomic treatment of ecotypic variation. In *Function and Taxonomic Importance*, Cain, A.J. (ed.), 87–112. Systematics Association, London.

200   Heywood, V.H. (1963). The 'species aggregate' in theory and practice. *Regnum veg.*, **27**, 26–37.

201 Heywood, V.H. (1971a). Preservation of the European flora. The taxonomist's role. *Bull. Jard. bot. nat. Belg.*, **41**, 153–166.
202 Heywood, V.H. (1971b). *Scanning Electron Microscopy: Systematic and Evolutionary Applications*. Academic Press, London and New York.
203 Heywood, V. H. (1973). Ecological data in practical taxonomy. In *Taxonomy and Ecology*, 329–347. Academic Press, London and New York.
204 Heywood, V.H. (1974). Systematics—the stone of Sisyphus. *Biol. J. Linn. Soc.*, **6**, 169–178.
205 Heywood, V.H. (1978). Flora Europaea. Notulae systematicae ad Floram Europaeam spectantes, 20. *Bot. J. Linn. Soc.*, **76**, 297–384.
206 Hibberd, D.J. (1976). The ultrastructure and taxonomy of the Chrysophyceae and Prymnesiophyceae (Haptophyceae): a survey with some new observations on the ultrastructure of the Chrysophyceae. *Bot. J. Linn. Soc.*, **72**, 55–80.
207 Hickey, L.J. (1973). Classification of the architecture of dicotyledonous leaves. *Am. J. Bot.*, **60**, 17–33.
208 Hickey, L.J. and Wolfe, J.A. (1975). The bases of angiosperm phylogeny: Vegetative morphology. *Ann. Mo. bot. Gdn*, **62**, 538–589.
209 Hollings, E. and Stace, C.A. (1974). Karyotype variation and evolution in the *Vicia sativa* aggregate. *New Phytol.*, **73**, 195–208.
210 Hollings, E. and Stace, C.A. (1978). Morphological variation in the *Vicia sativa* L. aggregate. *Watsonia*, **12**, 1–14.
211 Holmes, N.T.H. (1979). A guide to identification of Batrachium *Ranunculus* species of Britain. *Nature Conservancy Council, Chief Scientist's Team Notes*, **14**.
212 Hosaka, K. (1986). Who is the mother of the potato?—restriction endonuclease analysis of chloroplast DNA of cultivated potatoes. *Theor. Appl. Genet.*, **72**, 606–618.
213 Hultén, E. (1950). *Atlas of the Distribution of Vascular Plants in Northwestern Europe*. Generalstabens Litografiska Anstalts Förlag, Stockholm.
214 Humphries, C.J. and Funk, V.A. (1984). Cladistic methodology. In *Current Concepts in Plant Taxonomy*, Heywood, V.H. and Moore, D.M. (eds.), 323–362. Academic Press, London.
215 Huneck, S. (1983). Chemistry and biochemistry of bryophytes. In *New Manual of Bryology*, Schuster, R.M. (ed.), **1**, 1–116. Hattori Botanical Laboratory, Nichinan, Japan.
216 Hurka, H. (1980). Enzymes as a taxonomic tool: a botanist's view. In *Chemosystematics: Principles and Practice*, Bisby, F.A., Vaughan, J.G. and Wright, C.A. (eds), 103–121, Academic Press, London.
217 Hutchinson, J. (1926). *The Families of Flowering Plants*, **1**. Dicotyledons. Macmillan, London.
218 Hutchinson, J. (1973). *The Families of Flowering Plants*. 3rd edition. Oxford University Press.
219 Hutchinson, J.B., Silow, R.A. and Stephens, S.G. (1947). *The Evolution of Gossypium and the Differentiation of the Cultivated Cottons*. Oxford University Press.
220 Huxley, J.S. (1955). Morphism and evolution. *Heredity, Lond.*, **9**, 1–52.
221 Huxley, J.S. (1959). Clades and grades. In *Function and Taxonomic Importance*, Cain, A.J. (ed.), 21–22. Systematics Association, London.
222 Inamdar, J.A., Mohan, J.S.S. and Subramanian, R.B. (1986). Stomatal classifications—A review. *Feddes Repert.*, **97**, 147–160.
223 Ingram, R. (1967). On the identity of the Irish populations of *Sisyrinchium*. *Watsonia*, **6**, 283–289.

224   Ingrouille, M.J. and Stace, C.A. (1985). Pattern of variation of agamospermous *Limonium* (Plumbaginaceae) in the British Isles. *Nordic J. Bot.*, **5**, 113–125.
225   Jackson, R.C. (1971). The karyotype in systematics. *A. Rev. Ecol. Syst.*, **2**, 327–368.
226   Jain, S.K. (1976). The evolution of inbreeding in plants. *A. Rev. Ecol. Syst.*, **7**, 469–495.
227   Jalas, J. and Suominen, J. (1972). *Atlas Florae Europaeae*, **1**. *Pteridophyta (Psilotaceae to Azollaceae)*. Societas Biologica Fennica Vanamo, Helsinki.
228   Jeffrey, C. (1977). *Biological Nomenclature*. 2nd edition. Edward Arnold, London.
229   Jeffrey, C. (1982). Kingdoms, codes and classification. *Kew Bull.*, **37**, 403–416.
230   Jeffreys, A.J. (1985). Hypervariable 'minisatellite' regions in human DNA. *Nature, Lond.*, **314**, 67–73.
231   Jensen, U. (1968). Serologische Beitrage zur Systematik der Ranunculaceae. *Bot. Jb.*, **88**, 204–268.
232   Jensen, U. (1974). The interpretation of comparative serological results. In *Chemistry in Botanical Classification. Nobel Symposium*, **25**, Bendz, G. and Santesson, J. (eds), 217–227.
233   Jensen, U. and Fairbrothers, D.E. (eds) (1983). *Proteins and Nucleic Acids in Plant Systematics*. Springer-Verlag, Berlin.
234   Johnson, B.L. (1972). Seed protein profiles and the origin of the hexaploid wheats. *Am. J. Bot.*, **59**, 952–960.
235   Johri, B.M. (ed.) (1984). *Embryology of Angiosperms*. Springer-Verlag, Berlin.
236   Jones, B.M.G. and Newton, L.E. (1970). The status of *Puccinellia pseudodistans* (Crép.) Jansen & Wachter in Great Britain. *Watsonia*, **8**, 17–26.
237   Jones, D.A. (1973). Co-evolution and cyanogenesis. In *Taxonomy and Ecology*, Heywood, V.H. (ed.), 213–242. Academic Press, London and New York.
238   Jones, J.D.G. and Flavell, R.B. (1982). The structure, amount, and chromosomal localization of defined repeated DNA sequences in species of the genus *Secale*. *Chromosoma*, **86**, 613–641.
239   Jones, K. (1978). Aspects of chromosome evolution in higher plants. *Adv. bot. Res.*, **6**, 119–194.
240   Jones, K. and Jopling, C. (1972). Chromosomes and the classification of the Commelinaceae. *Bot. J. Linn. Soc.*, **65**, 129–162.
241   Jordan, A. (1873). Remarques sur le fait de l'existence en société à l'état sauvage des espèces végétales affines. *Bull. Ass. fr. Avanc. Sci.*, **2**, Session Lyon.
242   Jorgensen, L.B. (1981). Myrosin cells and dilated cisternae of the endoplasmic reticulum in the order Capparales. *Nordic J. Bot.*, **1**, 433–445.
243   Jorgensen, R.A., Cuellar, R.E. and Thompson, W.F. (1982). Modes and tempos in the evolution of nuclear-encoded ribosomal RNA genes in legumes. *Carnegie Inst. Year Book*, **81**, 98–101.
244   Jurgens, N. (1985). Konvergente Evolution von Blatt- und Epidermismerkmalen bei blattsukkulenten Familien. *Ber. dt. bot. Ges.*, **98**, 425–446.
245   Kay, Q.O.N. (1978). The role of preferential and assortive pollination in the maintenance of flower colour polymorphisms. In *The Pollination of Flowers by Insects*, Richards, A.J. (ed.), 175–190. Academic Press, London and New York.
246   Kay, Q.O.N. (1984). Variation, polymorphism and gene-flow within species. In *Current Concepts in Plant Taxonomy*, Heywood, V.H. and Moore, D.M. (eds), 181–199. Academic Press, London.
247   Kelley, W.A. and Adams, R.P. (1977). Seasonal variation of isozymes in *Juniperus scopulorum:* systematic significance. *Am. J. Bot.*, **64**, 1092–1096.

248 Knobloch, I. W. (1972). Intergeneric hybridization in flowering plants. *Taxon*, **21**, 97–103.

249 Knobloch I.W. (1976). *Pteridophyte Hybrids. Publs of the Museum, Biol.*, **5(4)**. Michigan State University, East Lansing.

250 Knobloch I.W., Gibby, M. and Fraser-Jenkins, C. (1984). Recent advances in our knowledge of pteridophyte hybrids. *Taxon*, **33**, 256–270.

251 Kociolek, J.P. and Stoermer, E.F. (1986). Phylogenetic relationships and classification of monoraphid diatoms based on phenetic and cladistic methodologies. *Phycologia*, **25**, 297–303.

252 Kössel, H., Edwards, K., Fritzsche, E., Koch, W. and Schwarz, Z. (1983). Phylogenetic significance of nucleotide sequence analysis. In *Proteins and Nucleic Acids in Plant Systematics*, Jensen, U. and Fairbrothers, D.E. (eds), 36–57. Springer-Verlag, Berlin.

253 Kruckeberg, A.R. (1969). Ecological aspects of the systematics of plants. In *Systematic Biology*, Sibley, C.G. (ed.), 161–203. National Academy of Sciences, Washington D.C.

254 Lane, M.A. and Turner, B.L. (eds) (1985). The generic concept in Compositae: a symposium. *Taxon*, **34**, 5–88.

255 Lawrence, G.H.M. (1951). *Taxonomy of Vascular Plants*. Macmillan, New York.

256 Leenhouts, P.W. (1976). A conspectus of the genus *Allophylus* (Sapindaceae). The problem of the complex species. *Blumea*, **15**, 301–358.

257 Leone, C.A. (ed.) (1964). *Taxonomic Biochemistry and Serology*. Ronald Press, New York.

258 Levan, A., Fredga, K. and Sandberg, A.A. (1965). Nomenclature for centromeric position on chromosomes. *Hereditas*, **52**, 201–220.

259 Lewis, H. (1967). The taxonomic significance of autopolyploidy. *Taxon*, **16**, 267–271.

260 Lewis, W.H. (ed.) (1980). *Polyploidy. Biological Relevance*. Plenum Press, New York and London.

261 Linder, H.P. (1987). The evolutionary history of the Poales/Restionales—a hypothesis. *Kew Bull.*, **42**, 297–318.

262 Lousley, J.E. (1953). The recent influx of aliens into the British flora. In *The Changing Flora of Britain*, 140–159. Botanical Society of the British Isles, Oxford.

263 Löve, Á. (1951). Taxonomical evaluation of polyploids. *Caryologia*, **3**, 263–284.

264 Löve, Á. and Löve, D. (1974). *Cytotaxonomical Atlas of the Slovenian Flora*. Cramer, Königstein.

265 Löve, Á., Löve, D. and Pichi-Sermolli, R.E.G. (1977). *Cytotaxonomical Atlas of the Pteridophytes*. Cramer, Königstein.

266 Loveless, M.D. and Hamrick, J.L. (1984). Ecological determinants of genetic structure in plant populations. *Ann. Rev. Ecol. Syst.*, **15**, 65–95.

267 Mabry, T.J. (1976). Pigment dichotomy and DNA-RNA hybridization data for Centrospermous families. *Pl. Syst. Evol.*, **126**, 79–94.

268 Mabry, T.J. (1980). Betalains. In *Encyclopedia of Plant Physiology, 8. Secondary Plant Products*, Bell, E.A. and Charlwood, B.V. (eds), 513–533. Springer-Verlag, Berlin.

269 Mabry, T.J. and Behnke, H.-D. (1976). Evolution of Centrospermous families. *Pl. Syst. Evol.*, **126**, 1–106.

270 Margulis, L. (1981). *Symbiosis in Cell Evolution*. W. H. Freeman, San Francisco.

271 Marsden-Jones, E.M. and Turrill, W.B. (1957). *The Bladder Campions*. Ray Society, London.

272 Martin, P.G., Boulter, D. and Penny, D. (1985). Angiosperm phylogeny studied using sequences of five macromolecules. *Taxon*, **34**, 339–400.

273 Martin, P.G. and Dowd, J.M. (1986). A phylogenetic tree for some mono-cotyledons and gymnosperms derived from protein sequences. *Taxon*, **35**, 469–475.

274 Matthews, J.R. (1937). Geographical relationships of the British flora. *J. Ecol.*, **25**, 1–90.

275 Mayr, E. (1942). *Systematics and the Origin of Species*. Columbia University Press, New York.

276 McKelvey, S.D. and Sax, K. (1933). Taxonomic and cytological relationships of *Yucca* and *Agave*. *J. Arn. Arb.*, **14**, 76–81.

277 McMillan, C., Mabry, T.J. and Chavez, P.I. (1976). Experimental hybridization of *Xanthium strumarium* (Compositae) from Asia and America, II. Sesquiter-pene lactones of the F_1 hybrids. *Am. J. Bot.*, **63**, 317–323.

278 Meacham, C.A. (1981). A manual method for character compatibility analysis. *Taxon*, **30**, 591–600.

279 Meacham, C.A. (1984). The role of hypothesized direction of characters in the estimation of evolutionary history. *Taxon*, **33**, 26–38.

280 Meacham, C.A. (1986). More about directed characters: a reply to Donoghue and Maddison. *Taxon*, **35**, 538–540.

281 Meacham, C.A. and Duncan, T. (1987). The necessity of convex groups in biological classification. *Syst. Bot.*, **12**, 78–90.

282 Meeuse, A.D.J. (1963). Some phylogenetic aspects of the process of double fertilization. *Phytomorphol.*, **13**, 237–244.

283 Meeuse, A.D.J. (1973a). Anthecology and angiosperm evolution. In *Taxonomy and Ecology*, Heywood, V.H. (ed.), 189–200. Academic Press, London and New York.

284 Meeuse, A.D.J. (1973b). Co-evolution of plant hosts and their parasites as a taxonomic tool. In *Taxonomy and Ecology*, Heywood, V.H. (ed.), 289–316. Academic Press, London and New York.

285 Melchior, H. (1964). *A. Engler's Syllabus der Pflanzenfamilien*, **2**, 12th edition. Gebrüder Borntraeger, Berlin.

286 Melville, R. (1955). Morphological characters in the discrimination of species and hybrids. In *Species Studies in the British Flora*, Lousley, J.E. (ed.), 55–64. Botanical Society of the British Isles, London.

287 Melville, R. (1976). The terminology of leaf architecture. *Taxon*, **25**, 549–561.

288 Mensch, J.A. and Gillett, G.W. (1972). The experimental verification of natural hybridization between two taxa of Hawaiian *Bidens* (Asteraceae). *Brittonia*, **24**, 57–70.

289 Metcalfe, C.R. (1960). *Anatomy of the Monocotyledons*, I. *Gramineae*. Oxford University Press.

290 Metcalfe, C.R. (1963). Comparative anatomy as a modern botanical discipline, with special reference to recent advances in the systematic anatomy of mono-cotyledons. *Adv. bot. Res.*, **1**, 101–147.

291 Metcalfe, C.R. and Chalk, L. (1950). *Anatomy of the Dicotyledons*, **1** and **2**. Oxford University Press.

292 Metcalfe, C.R. and Chalk, L. (eds) (1979–1983). *Anatomy of the Dicotyledons*. 2nd edition. **1** (1979), **2** (1983). Clarendon Press, Oxford.

293 Michener, C.D. and Sokal, R.R. (1957). A quantitative approach to a problem in classification. *Evolution, Lancaster, Pa.*, **11**, 130–162.

294 Mirov, N.T. (1967). *The Genus Pinus*. Ronald Press, New York.

295 Moore, D.M. (1972). Connections between cool temperate floras, with particu-lar reference to southern South America. In *Taxonomy, Phytogeography and*

Evolution, Valentine, D.H. (ed.), 115–138. Academic Press, London and New York.

296 Moore, D.M. (1984). Taxonomy and geography. In *Current Concepts in Plant Taxonomy*, Heywood, V.H. and Moore, D.M. (eds), 219–234. Academic Press, London.

297 Moore, D.M. and Chater, A.O. (1971). Studies of bipolar disjunct species, I. *Carex. Bot. Notiser*, **124**, 317–334.

298 Morisset, P. (1978). Chromosome numbers in *Ononis* L. series *Vulgares* Širj. *Watsonia*, **12**, 145–153.

299 Morton, J.K. (1972). Phytogeography of the West African mountains. In *Taxonomy, Phytogeography and Evolution*, Valentine, D.H. (ed.), 221–236. Academic Press, London.

300 Muller, J. (1981). Fossil pollen records of extant angiosperms. *Bot. Rev.*, **47**, 1–142.

301 Munz, P.A. (1928–1935). Studies in Onagraceae, 1–9. *Am. J. Bot.*, **15**, 223–240 (1928); *Bot. Gaz.*, **85**, 243–270 (1928); *Am. J. Bot.*, **16**, 246–257 (1929); **16**, 702–715 (1929); **17**, 358–370 (1930); **18**, 309–327 (1931); **18**, 728–738 (1931); **19**, 755–778 (1932); **22**, 645–663 (1935).

302 Nelson, A.P. (1965). Taxonomic and evolutionary implications of lawn races in *Prunella vulgaris* (Labiatae). *Brittonia*, **17**, 160–174.

303 Nelson, C.H. and van Horn, G.S. (1975). A new simplified method for constructing Wagner networks and the cladistics of *Pentachaeta* (Compositae, Astereae). *Brittonia*, **27**, 362–372.

304 Nelson, G.J. and Platnick, N.I. (1981). *Systematics and Biogeography, Cladistics and Vicariance*. Columbia University Press, New York.

305 New, J.K. (1978). Change and stability of clines in *Spergula arvensis* L. (corn spurrey) after 20 years. *Watsonia*, **12**, 137–143.

306 Newton, A. (1975). *Rubus* L. In *Hybridization and the Flora of the British Isles*, Stace, C.A. (ed.), 200–206. Academic Press, London and New York.

307 Newton, A. (1980). Progress in British *Rubus* studies. *Watsonia*, **13**, 35–40.

308 Newton, M.E. (1983). Cytology of the Hepaticae and Anthocerotae. In *New Manual of Bryology*, Schuster, R.M. (ed.), 117–148. Hattori Botanical Laboratory, Nichinan, Japan.

309 Newton, M.E. (1984). The cytogenetics of bryophytes. In *The Experimental Biology of Bryophytes*, Dyer, A.F. and Duckett, J.G. (eds), 65–96. Academic Press, London.

310 Nilsson, N.H. (1954). Über Hochkomplexe Bastardverbindungen in der Gattung *Salix. Hereditas*, **40**, 517–522.

311 Nilsson, S. (ed.) (1973→). World pollen and spore flora, 1→. *Grana, Int. J. Palynol., Suppl.*

312 Okafor, J.C. (1978). Development of forest tree crops for food supplies in Nigeria. *For. Ecol. Management*, **1**, 235–247.

313 Palser, B.F. (1975). The bases of angiosperm phylogeny. Embryology. *Ann. Mo. bot. Gdn*, **62**, 621–646.

314 Pankhurst, R.J. (ed.) (1975). *Biological Identification with Computers*. Academic Press, London and New York.

315 Pankhurst, R.J. (1978). *Biological Identification*. Edward Arnold, London.

316 Pant, D.D. (1965). On the ontogeny of stomata and other homologous structures. *Pl. Sci. Ser.*, **1**, 1–24.

317 Pant, D.D. and Kidwai, P.F. (1964). On the diversity in the development and organization of stomata in *Phyla nodiflora* Michx. *Curr. Sci.*, **33**, 653–654.

318 Parker, P.F. (1978). The classification of crop plants. In *Essays in Plant*

Taxonomy, Street, H.E. (ed.), 97–124. Academic Press, London and New York.

319   Parker, P.F. (1986). The classification of cultivated plants—problems and prospects. In *Infraspecific Classification of Wild and Cultivated Plants*, Styles, B.T. (ed.), 99–114. Clarendon Press, Oxford.

320   Patel, V.C., Skvarla, J.J. and Raven, P.H. (1985). Pollen characters in relation to the delimitation of Myrtales. *Ann. Mo. bot. Gdn*, **71**, 858–969.

321   Patterson, C. (1980). Cladistics. *Biologist*, **27**, 234–240.

322   Payne, W.W. (1978). A glossary of plant hair terminology. *Brittonia*, **30**, 239–255.

323   Perring, F.H. and Walters, S.M. (eds) (1962). *Atlas of the British Flora*. Botanical Society of the British Isles, London.

324   Phillips, E.W.J. (1948). *Identification of Softwoods by their Microscopic Structure. Forest Products Research Bulletin*, **22**. Her Majesty's Stationery Office, London.

325   Pichi Sermolli, R.E.G. (1973). Historical review of the higher classification of the Filicopsida. In *The Phylogeny and Classification of the Ferns, Bot. J. Linn. Soc.*, **67**, Suppl. **1**, Jermy, A.C., Crabbe, J.A. and Thomas, B.A. (eds), 11–40.

326   Platnick, N.I. (1979). Philosophy and the transformation of cladistics. *Syst. Zool.*, **28**, 537–546.

327   Poelt, J. (1973). Appendix A, Classification. In *The Lichens*, Ahmadjian, V. and Hale, M.E. (eds), 599–632. Academic Press, London and New York.

328   Porter, L.J. (1981). Geographical races of *Conocephalum* (Marchantiales) as defined by flavonoid chemistry. *Taxon*, **30**, 739–748.

329   Prance, G.T. (1978). Floristic inventory of the tropics: Where do we stand? *Ann. Mo. bot. Gdn*, **64**, 659–684. Note correction in *Ann. Mo. bot. Gdn*, **65**, i–ii (1978).

330   Prance, G.T. (1984). Completing the inventory. In *Current Concepts in Plant Taxonomy*, Heywood, V.H. and Moore, D.M. (eds), 365–396. Academic Press, London.

331   Pritchard, N.M. (1961). *Gentianella* in Britain, 3. *Gentianella germanica* (Willd.) Börner. *Watsonia*, **4**, 290–303.

332   Proctor, M.C.F. (1978). Insect pollination syndromes in an evolutionary and ecosystemic context. In *The Pollination of Flowers by Insects*, Richards, A.J. (ed.), 105–116. Academic Press, London and New York.

333   Prus-Głowacki, W. (1983). Serological investigation of a hybrid swarm population of *Pinus sylvestris* L. × *P. mugo* Turra, and the antigenic differentiation of *Pinus sylvestris* L. in Sweden. In *Proteins and Nucleic Acids in Plant Systematics*, Jensen, U. and Fairbrothers, D.E. (eds), 352–361. Springer-Verlag, Berlin.

334   Punt, W. (ed.) (1976→). *The Northwest European Pollen Flora*, 1→. Elsevier, Amsterdam.

335   Radford, A.E., Dickison, W.C., Massey, J.R. and Bell, C.R. (1974). *Vascular Plant Systematics*. Harper & Row, New York and London.

336   Rahn, K. (1983). Phenetic and phylogenetic studies based on measurements of *Plantago* ser. *Brasilienses. Nordic J. Bot.*, **3**, 319–329.

337   Ramayya, N. (1972). Classification and phylogeny of the trichomes of angiosperms. In *Research Trends in Plant Taxonomy—K. A. Chowdhury Commemoration Volume*, Ghouse, A.K.M. and Yunus, M. (eds), 91–102. Tata McGraw-Hill, New Delhi.

338   Raven, J.E. (1977). The evolution of vascular land plants in relation to supracellular transport processes. *Adv. bot. Res.*, **15**, 153–219.

339   Raven, P.H. (1963a). *Circaea* in the British Isles. *Watsonia*, **5**, 262–272.

340 Raven, P.H. (1963b). Amphitropical relationships in the floras of North and South America. *Q. Rev. Biol.*, **38**, 151–177.

341 Raven, P.H. (1974). Nomenclature proposals to the Leningrad Congress. *Taxon*, **23**, 828–833.

342 Raven, P.H. (1975). The bases of angiosperm phylogeny: Cytology. *Ann. Mo. bot. Gdn*, **62**, 724–764.

343 Raven, P.H. (1976a). Systematic botany and plant population biology. *Syst. Bot.*, **1**, 284–316.

344 Raven, P.H.' (1976b). Generic and sectional delimitation in Onagraceae, tribe Epilobieae. *Ann. Mo. bot. Gdn*, **63**, 326–340.

345 Raven, P.H. (1978). Perspectives in tropical botany: Concluding remarks. *Ann. Mo. bot. Gdn*, **64**, 746–748.

346 Raven, P.H. (1979). Plate tectonics and southern hemisphere biogeography. In *Tropical Botany*, Larsen, K. and Holm-Nielsen, L.B. (eds), 3–24. Academic Press, London.

347 Raven, P.H. and Axelrod, D.I. (1978). Origin and relationships of the California flora. *Univ. Calif. Publs Bot.*, **72**.

348 Raven, P.H., Raven, T.E. and West, K.R. (1976). *The Genus* Epilobium *(Onagraceae) in Australasia: A Systematic and Evolutionary Study. Bulletin* **216**. New Zealand D.S.I.R., Christchurch.

349 Reed, C.F. (1953). Index Isoetales. *Bol. Soc. broteriana*, Sér. 2, **27**, 5–72.

350 Reed, C.F. (1966a). Index Psilotales. *Bol. Soc. broteriana*, Sér. 2, **40**, 71–96.

351 Reed, C.F. (1966b). Index Selaginellarum. *Mem. Soc. broteriana*, **18**, 5–287.

352 Reed, C.F. (1971). *Index to Equisetophyta, Part II. Extantes. Index Equisetorum*. Reed Herbarium, Baltimore.

353 Rees, H. (1974). B chromosomes. *Sci. Progr. (Oxf.)*, **61**, 535–554.

354 Regan, C.T. (1925). Organic evolution: facts and theories. *Br. Assoc. Adv. Sci., Rep. 93rd Meeting*, 75–86.

355 Ribereau-Gayon, P. (1972). *Plant Phenolics*. Oliver & Boyd, Edinburgh and London.

356 Richards, A.J. (1973). The origin of *Taraxacum* agamospecies. *Bot. J. Linn. Soc.*, **66**, 189–211.

357 Richards, A.J. (1986). *Plant Breeding Systems*. Allen & Unwin, London.

358 Richards, P.W. (1978). The taxonomy of bryophytes. In *Essays in Plant Taxonomy*, Street, H.E. (ed.), 177–209. Academic Press, London and New York.

359 Richardson, I.B.K. (1978). Endemic taxa and the taxonomist. In *Essays in Plant Taxonomy*, Street, H.E. (ed.), 245–262. Academic Press, London and New York.

360 Richens, R.H. and Jeffers, J.N.R. (1975). Multivariate analysis of the elms of northern France, I. Variation within France. *Silvae Genet.*, **24(5–6)**, 129–200.

361 Ride, W.D.L. and Younès, T. (eds) (1986). *Biological Nomenclature Today*. IRL Press, Oxford.

362 Riley, R. and Chapman, V. (1958). Genetic control of the cytologically diploid behaviour of hexaploid wheat. *Nature, Lond.*, **183**, 713–715.

363 Robson, N.K.B., Cutler, D.F. and Gregory, M. (eds) (1970). *New Research in Plant Anatomy*. Academic Press, London and New York.

364 Romeike, A. (1978). Tropane alkaloids—occurrence and systematic importance in angiosperms. *Bot. Notiser*, **131**, 85–96.

365 Rose, F. (1972). Floristic connections between southeast England and north France. In *Taxonomy, Phytogeography and Evolution*, Valentine, D.H. (ed.), 363–379. Academic Press, London and New York.

366   Ross, R. and Sims, P.A. (1973). Observations on family and generic limits in the Centrales. *Nova Hedwigia*, Beihefte **45**, 97–128.

367   Royal Botanic Gardens, Kew (1980). *Draft Index of Author Abbreviations compiled at The Herbarium, Royal Botanic Gardens, Kew*. H.M.S.O., London.

368   Rushton, B.S. (1978). *Quercus robur* L. and *Quercus petraea* (Matt.) Liebl.: a multivariate approach to the hybrid problem, 1. Data acquisition, analysis and interpretation. *Watsonia*, **12**, 81–101.

369   Saether, O.A. (1986). The myth of objectivity—post-Hennigian deviations. *Cladistics*, **2**, 1–13.

370   Sahasrabudhe, S. and Stace, C.A. (1974). Developmental and structural variation in the trichomes and stomata of some Gesneriaceae. *New Botanist*, **1**, 46–62.

371   Savile, D.B.O. (1954). The fungi as aids in the taxonomy of flowering plants. *Science, N.Y.*, **120**, 583–585.

372   Scannell, M.J.P. (1975). *Juncus planifolius* R. Br. in Ireland. *Ir. Nat. J.*, **17**, 308–309.

373   Schlichting, C.D. and Levin, D.A. (1986). Phenotypic plasticity: an evolving plant character. *Biol. J. Linn. Soc.*, **29**, 37–47.

374   Schweizer, D. and Ehrendorfer, F. (1976). Giemsa-banded karyotypes, systematics, and evolution in *Anacyclus* (Asteraceae–Anthemideae). *Pl. Syst. Evol.*, **126**, 107–148.

375   Sharma, A.K. and Sharma, A. (1980). *Chromosome Techniques*. Butterworths, London.

376   Shibata, S. (1974). Some aspects of lichen chemotaxonomy. In *Chemistry in Botanical Classification. Nobel Symposium*, **25**, Bendz, G. and Santesson, J. (eds), 241–249. Academic Press, London and New York.

377   Shimwell, D.W. (1971). *The Description and Classification of Vegetation*. Sidgwick & Jackson, London.

378   Shull, G.H. (1929). Species hybridizations among old and new species of Shepherd's Purse. In *Proc. International Congress of Plant Sciences, Ithaca, New York, August 16–23, 1926*, **1**, Duggar, B.M. (ed.), 832–888. Menasha, Wisconsin.

379   Sibley, C.G. (1962). The comparative morphology of protein molecules as data for classification. *Syst. Zool.*, **11**, 108–118.

380   Sinker, C.A. (1975). A lateral key to common grasses. *Bull. Shropshire Conserv. Trust*, **34**, 11–18.

381   Smith, A.J.E. (1978). Cytogenetics, biosystematics and evolution in the Bryophyta. *Adv. bot. Res.*, **6**, 195–276.

382   Smith, A.J.E. (1979). Towards an experimental approach to bryophyte taxonomy. In *Bryophyte Systematics*, Clarke, G.C.S. and Duckett, J.G. (eds), 195–206. Academic Press, London.

383   Smith, D.M. and Levin, D.A. (1963). A chromatographic study of reticulate evolution in the Appalachian *Asplenium* complex. *Am. J. Bot.*, **50**, 952–958.

384   Smith, P.M. (1972). Serology and species relationships in annual bromes (*Bromus* L. sect. *Bromus*). *Ann. Bot.*, **36**, 1–30.

385   Smith P. M. (1976). *The Chemotaxonomy of Plants*. Edward Arnold, London.

386   Smith, P.M. (1978). Chemical evidence in plant taxonomy. In *Essays in Plant Taxonomy*, Street, H.E. (ed.), 19–38. Academic Press, London and New York.

387   Smith, P.M. (1983). Proteins, mimicry and microevolution in grasses. In *Proteins and Nucleic Acids in Plant Systematics*, Jensen, U. and Fairbrothers, D.E. (eds), 311–323. Springer-Verlag, Berlin.

388   Snaydon, R.W. (1973). Ecological factors, genetic variation and speciation in plants. In *Taxonomy and Ecology*, Heywood, V.H. (ed.), 1–29. Academic Press, London and New York.

389 Snaydon, R.W. (1984). Infraspecific variation and its taxonomic implications. In *Current Concepts in Plant Taxonomy*, Heywood, V.H. and Moore, D.M. (eds), 203–218. Academic Press, London.

390 Sneath, P.H.A. (1957a). Some thoughts on bacterial classification. *J. gen. Microbiol.*, **17**, 184–200.

391 Sneath, P.H.A. (1957b). The application of computers to taxonomy. *J. gen. Microbiol.*, **17**, 201–226.

392 Sneath, P.H.A. (1972). Computer taxonomy. In *Methods in Microbiology*, **7A**, Norris, J.R. and Ribbons, D.W. (eds), 29–98. Academic Press, London and New York.

393 Sneath, P.H.A. and Sokal, R.R. (1973). *Numerical Taxonomy*. W. H. Freeman, San Francisco.

394 Sokal, R.R. and Sneath, P.H.A. (1963). *Principles of Numerical Taxonomy*. W. H. Freeman, San Francisco.

395 Solbrig, O.T. (1968). Fertility, sterilty and the species problem. In *Modern Methods in Plant Taxonomy*, Heywood, V.H. (ed.), 77–96. Academic Press, London and New York.

396 Solbrig, O.T. (1970a). *Principles and Methods of Plant Biosystematics*. Macmillan, London.

397 Solbrig, O.T. (1970b). The phylogeny of *Gutierrezia:* An eclectic approach. *Brittonia*, **22**, 217–229.

398 Sporne, K.R. (1973). The survival of archaic dicotyledons in tropical rainforests. *New Phytol.*, **72**, 1175–1184.

399 Sporne K.R. (1974). *The Morphology of Angiosperms*. Hutchinson University Library, London.

400 Sporne, K.R. (1977). Some problems associated with character correlations. *Pl. Syst. Evol.*, Suppl. **1**, 33–51.

401 Stace, C.A. (1961). Some studies in *Calystegia:* Compatibility and hybridization in *C. sepium* and *C. silvatica*. *Watsonia*, **5**, 88–105.

402 Stace, C.A. (1965a). Cuticular characters as an aid to plant taxonomy. *Bull. Br. Mus. (Nat. Hist.), Bot.*, **4**, 3–78.

403 Stace, C.A. (1965b). The significance of the leaf epidermis in the taxonomy of the Combretaceae, I. A general review of tribal, generic and specific characters. *J. Linn. Soc. (Bot.)*, **59**, 229–252.

404 Stace, C.A. (1973). Chromosome numbers in the British species of *Calystegia* and *Convolvulus*. *Watsonia*, **9**, 363–367.

405 Stace, C.A. (ed.) (1975). *Hybridization and the Flora of the British Isles*. Academic Press, London and New York.

406 Stace, C.A. (1976). The study of infraspecific variation. *Curr. Adv. Pl. Sci.*, **8(4)**, Commentary 23.

407 Stace, C.A. (1978). Breeding systems, variation patterns and species delimitation. In *Essays in Plant Taxonomy*, Street, H.E. (ed.), 57–78. Academic Press, London and New York.

408 Stace, C.A. (1981). The significance of the leaf epidermis in the taxonomy of the Combretaceae: conclusion. *Bot. J. Linn. Soc.*, **81**, 327–339.

409 Stace, C.A. (1984). The taxonomic importance of the leaf surface. In *Current Concepts in Plant Taxonomy*, Heywood, V.H. and Moore, D.M. (eds), 67–94. Academic Press, London.

410 Stace, C.A. (1986). The present and future infraspecific classification of wild plants. In *Infraspecific Classification of Wild and Cultivated Plants*, Styles, B.T. (ed.), 9–20. Clarendon Press, Oxford.

411   Stace, C.A. and Auquier, P. (1978). Taxonomy and variation of *Vulpia ciliata* Dumort. *Bot. J. Linn. Soc.*, **77**, 107–112.

412   Stace, C.A. and Cotton, R. (1976). Nomenclature, comparison and distribution of *Vulpia membranacea* (L.) Dumort. and *V. fasciculata* (Forskål) Samp. *Watsonia*, **11**, 117–123.

413   Stafleu, F.A. (1971). *Linnaeus and the Linnaeans*. International Association for Plant Taxonomy, Utrecht.

414   Stearn, W.T. (1951). Mapping the distribution of species. In *The Study of the Distribution of British Plants*, Lousley, J.E. (ed.), 48–64. Botanical Society of the British Isles, Oxford.

415   Stearn, W.T. (1957–1959). An introduction to the *Species Plantarum* and cognate botanical works of Carl Linnaeus. In *Species Plantarum. A Facsimile of the First Edition 1753*. Ray Society, London.

416   Stearn, W.T. (1973). Introduction to Facsimile Reprint of Ray, J. *Synopsis Methodica Stirpium Britannicarum Editio Tertia 1724*. Ray Society, London.

417   Stearn, W.T. (1983). *Botanical Latin*. 3rd edition. David and Charles, Newton Abbot.

418   Stebbins, G.L. (1950). *Variation and Evolution in Plants*. Columbia University Press, New York and London.

419   Stebbins, G.L. (1959). The role of hybridization in evolution. *Proc. Am. phil. Soc.*, **103**, 231–251.

420   Stebbins, G.L. (1971). *Chromosomal Evolution in Higher Plants*. Edward Arnold, London.

421   Stebbins, G.L. (1972a). Ecological distribution of centers of major adaptive radiation in angiosperms. In *Taxonomy, Phytogeography and Evolution*, Valentine, D.H. (ed.), 7–34. Academic Press, London and New York.

422   Stebbins, G.L. (1972b). Research on the evolution of higher plants: Problems and prospects. *Can. J. Genet. Cytol.*, **14**, 453–462.

423   Stebbins, G.L. and Major, J. (1965). Endemism and speciation in the Californian flora. *Ecol. Monogr.*, **35**, 1–35.

424   Steinbrück, G., Schlegel, M., Dahlström, I. and Röttger, B. (1986). Characterization of interspecific hybrids between *Orchis mascula* and *O. pallens* by enzyme electrophoresis. *Pl. Syst. Evol.*, **153**, 229–241.

425   Stewart, K.D. and Mattox, K.R. (1975). Comparative cytology, evolution and classification of the green algae, with some consideration of the origin of other organisms with chlorophylls A and B. *Bot. Rev.*, **41**, 104–135.

426   Strid, A. (1970). Studies in the Aegean flora, XVI. Biosystematics of the *Nigella arvensis* complex with special reference to the problem of non-adaptive radiation. *Op. bot. Soc. bot. Lund.*, **28**, 1–169.

427   Strid, A. (1972). Some evolutionary and phytogeographical problems in the Aegean. In *Taxonomy, Phytogeography and Evolution*, Valentine, D.H. (ed.), 289–300. Academic Press, London and New York.

428   Stutz, H.C. and Thomas, L.K. (1964). Hybridization and introgression in *Cowania* and *Purshia*. *Evolution, Lancaster, Pa.*, **18**, 183–195.

429   Styles, B.T. (ed.) (1986). *Infraspecific Classification of Wild and Cultivated Plants*. Clarendon Press, Oxford.

430   Suire, C. and Asakawa, Y. (1979). Chemotaxonomy of bryophytes: a survey. In *Bryophyte Systematics*, Clarke, G.C.S. and Duckett, J.G. (eds), 447–477. Academic Press, London.

431   Swain, T. (ed.) (1963). *Chemical Plant Taxonomy*. Academic Press, London and New York.

432   Swain, T. (ed.) (1966). *Comparative Phytochemistry*. Academic Press, London and New York.

433 Sybenga, J. (1975). *Meiotic Configurations*. Springer-Verlag, Berlin.
434 Sytsma, K.J. and Gottlieb, L.D. (1986). Chloroplast DNA evolution and phylogenetic relationships in *Clarkia* section *Peripetasma* (Onagraceae). *Evolution, Lancaster, Pa.*, **40**, 1248–1261.
435 Szweykowski, J. and Krzakowa, M. (1979). Variation of four enzyme systems in Polish populations of *Conocephalum conicum*. *Bull. Acad. Pol. Sci.*, Ser. Sci. Biol., **27**, 37–42.
436 Takhtajan, A. (1987). *Sistema Magnoliophytov*. Nauka, Leningrad, (see *Taxon*, **37**, 422–424 (1988) for summary).
437 Thorne, R.F. (1973). The 'Amentiferae' or Hamamelidae as an artificial group: A summary statement. *Brittonia*, **25**, 395–405.
438 Thorne, R.F. (1983). Proposed new realignments in the angiosperms. *Nordic J. Bot.*, **3**, 85–117.
439 Tomlinson, P.B. (1984). Vegetative morphology—Some enigmas in relation to plant systematics. In *Current Concepts in Plant Taxonomy*, Heywood, V.H. and Moore, D.M. (eds), 49–66. Academic Press, London.
440 Turesson, G. (1922). The genotypical response of the plant species to the habitat. *Hereditas*, **3**, 211–350.
441 Turrill, W.B. (1935). The investigation of plant species. *Proc. Linn. Soc. Lond.*, **147**, 104–105.
442 Turrill, W.B. (1938). The expansion of taxonomy with special reference to the Spermatophyta. *Biol. Rev.*, **13**, 342–373.
443 Turrill, W.B. (1942). Taxonomy and phylogeny. *Bot. Rev.*, **8**, 247–270, 473–532, 655–707.
444 Ubsdell, R.A.E. (1979). Studies on variation and evolution in *Centaurium erythraea* Rafn and *C. littorale* (D. Turner) Gilmour in the British Isles, 3. Breeding systems, floral biology and general discussion. *Watsonia*, **12**, 225–232.
445 Urbanska, K.M. (1985). Some life history strategies and population structure in asexually reproducing plants. *Bot. Helv.*, **95**, 81–97.
446 Valentine, D.H. (1949). The units of experimental taxonomy. *Acta biotheor.*, **9**, 75–88.
447 Valentine, D.H. (1975). The taxonomic treatment of polymorphic variation. *Watsonia*, **10**, 385–390.
448 Valentine, D.H. (1978). Ecological criteria in plant taxonomy. In *Essays in Plant Taxonomy*, Street, H.E. (ed.), 1–18. Academic Press, London and New York.
449 Valentine, D.H. and Löve, Á. (1958). Taxonomic and biosystematic categories. *Brittonia*, **10**, 153–166.
450 Van Der Maesen, L.J.G. (ed.) (1986). *First International Symposium on Taxonomy of Cultivated Plants. Acta Horticulturae*, **182**. International Society for Horticultural Science, Wageningen.
451 Van Steenis, C.G.G.J. (1957). Specific and infraspecific delimitation. In *Flora Malesiana*, Ser. I, **5**, clxvii–ccxxxiv. Noordhoff, Djakarta.
452 Van Steenis, C.G.G.J. (1978). The doubtful virtue of splitting families. *Bothalia*, **12**, 425–427.
453 Vaughan, J.G. (1983). The use of seed proteins in taxonomy and phylogeny. In *Seed Proteins*, Daussant, J., Mossé, J. and Vaughan, J. (eds), 135–153. Academic Press, London.
454 Vavilov, N.I. (1951). *The Origin, Variation, Immunity and Breeding of Cultivated Plants*. Chronica Botanica Co., Waltham.
455 Wagner, W.H. (1964). Paraphyses: Filicineae. *Taxon*, **13**, 56–64.
456 Wagner, W.H. (1980). Origin and philosophy of the groundplan-divergence method of cladistics. *Syst. Bot.*, **5**, 173–193.
457 Wagner, W.H. (1983). Reticulistics: the recognition of hybrids and their role in

cladistics and classification. In *Advances in Cladistics*, **2**, Platnick, N.I. and Funk, V.A. (eds), 63–79. Columbia University Press, New York.

458 Walker, J.W. and Doyle, J.A. (1975). The bases of angiosperm phylogeny: Palynology. *Ann. Mo. bot. Gdn*, **62**, 664–723.

459 Walker, T.G. (1966). Apomixis and vegetative reproduction in ferns. In *Reproductive Biology and Taxonomy of Vascular Plants*, Hawkes, J.G. (ed.), 152–161. Pergamon Press, Oxford and London.

460 Walters, S.M. (1964). *Montia* L. In *Flora Europaea*, **1**, Tutin, T.G., Heywood, V.H., Burges, N.A., Valentine, D.H., Walters, S.M. and Webb, D.A. (eds), 114–115, Cambridge University Press.

461 Walters, S.M. (1966). The taxonomic implications of apomixis. In *Reproductive Biology and Taxonomy of Vascular Plants*, Hawkes, J.G. (ed.), 162–168. Pergamon Press, Oxford and London.

462 Walters, S.M. (1978). British endemics. In *Essays in Plant Taxonomy*, Street, H.E. (ed.), 263–274. Academic Press, London and New York.

463 Walters, S.M. (1986). The name of the rose: a review of ideas on the European bias in angiosperm classification. *New Phytol.*, **104**, 527–546.

464 Wanntorp, H.-E. (1983). Reticulated cladograms and the identification of hybrid taxa. In *Advances in Cladistics*, **2**, Platnick, N.I. and Funk, V.A. (eds), 81–88. Columbia University Press, New York.

465 Warburg, E.F. (1962). *Sorbus* L. In *Flora of the British Isles*, 2nd edition, Clapham A.R., Tutin, T.G. and Warburg, E.F., 423–437. Cambridge University Press.

466 Watson, L. and Dallwitz, M.J. (1985). *Australian Grass Genera. Anatomy, Morphology, Keys and Classification*. 2nd edition. Australian National University, Canberra.

467 Watson, L., Clifford, H.T. and Dallwitz, M.J. (1985). The classification of Poaceae: subfamilies and supertribes. *Austr. J. Bot.*, **33**, 433–484.

468 Watson, L., Dallwitz, M.J. and Johnston, C.R. (1986). Grass genera of the world: 728 detailed descriptions from an automated database. *Austr. J. Bot.*, **34**, 223–230.

469 Webb, D.A. (1978). Flora Europaea—a retrospect. *Taxon*, **27**, 3–14.

470 Weimarck, G. (1974). Population structure in *Circaea lutetiana*, *C. alpina* and *C. × intermedia* (Onagraceae) as revealed by thin-layer chromatographic patterns. In *Chemistry in Botanical Classification. Nobel Symposium*, **25**, Bendz, G. and Santesson, J. (eds), 287–292. Academic Press, London and New York.

471 Wenzel, W. and Hemleben, V. (1982). A comparative study of genomes in angiosperms. *Pl. Syst. Evol.*, **139**, 209–227.

472 Westfall, R.H., Glen, H.F. and Panagos, M.D. (1986). A new identification aid combining features of a polyclave and an analytical key. *Bot. J. Linn. Soc.*, **92**, 65–73.

473 White, M.J.D. (1973). *The Chromosomes*. 6th edition. Chapman & Hall, London.

474 Whitehouse, H.L.K. (1969). *Towards an Understanding of the Mechanism of Heredity*. 2nd edition. Edward Arnold, London.

475 Wildman, S.G. (1983). Polypeptide composition of Rubisco as an aid in studies of plant phylogeny. In *Proteins and Nucleic Acids in Plant Systematics*, Jensen, U. and Fairbrothers, D.E. (eds), 182–190. Springer-Verlag, Berlin.

476 Wiley, E.O. (1980). Phylogenetic systematics and vicariance biogeography. *Syst. Bot.*, **5**, 194–220.

477 Wiley, E.O. (1981). *Phylogenetics. The Theory and Practice of Phylogenetic Systematics*. John Wiley, New York.

478 Wilkinson, H.P. (1979). The plant surface (mainly leaf). In *Anatomy of the*

Dicotyledons, 2nd edition, Metcalfe, C.R. and Chalk, L. (eds), 97–165. Clarendon Press, Oxford.

479 Willemse, M.T.M. and Van Went, J.L. (1984). The female gametophyte. In *Embryology of Angiosperms*, Johri, B.M. (ed.), 159–196. Springer-Verlag, Berlin.

480 Willis, J.C. (1922). *Age and Area*. Cambridge University Press.

481 Wilmott, A.J. (1950). Systematic botany from Linnaeus to Darwin. In *Lectures on the Development of Taxonomy*, 33–45. Linnean Society of London, London.

482 Winge, O. (1940). Taxonomic and evolutionary studies in *Erophila* based on cytogenetic investigations. *C. R. Trav. Lab. Carlsberg,* Sér. Physiol., **24**, 41–74.

483 Yeo, P.F. (1975). *Euphrasia* L. In *Hybridization and the Flora of The British Isles*, Stace, C.A. (ed.), 373–381. Academic Press, London and New York.

484 Yeo, P.F. (1978). *Euphrasia:* a taxonomically critical group with normal sexual reproduction. In *Essays in Plant Taxonomy*, Street, H.E. (ed.), 143–162. Academic Press, London and New York.

485 Young, D.J. and Watson, L. (1970). The classification of the dicotyledons: A study of the upper levels of the hierarchy. *Austr. J. Bot.*, **18**, 387–433.

486 Zeven, A.C. and Zhukovsky, P.M. (1975). *Dictionary of Cultivated Plants and their Centres of Diversity*. Centre for Agricultural Publishing and Documentation, Wageningen.

487 Zielinski, R., Szweykowski, J. and Rutkowska, E. (1985). A further electrophoretic study of peroxidase isoenzyme variation in *Pellia epiphylla* (L.) Dum. from Poland, with special reference to the status of *Pellia borealis* Lorbeer. *Monogr. syst. Bot. Mo. bot. Gdn*, **11**, 199–210.

488 Favre-Duchartre, M. (1984). Homologies and phylogeny. In *Embryology of Angiosperms*, Johri, B.M. (ed.), 697–734. Springer-Verlag, Berlin.

489 Landolt, E. (1986–1987). The family of Lemnaceae—a monographic study, 1 and 2. In Biosystematic investigations in the family of duckweeds (Lemnaceae), 2 and 4. *Veröff. geobot. Inst., Zürich*, **71** and **95**.

490 Sporne, K.R. (1980). A re-investigation of character correlations among dicotyledons. *New Phytol.*, **85**, 419–449.

INDEX